WHAT IS QUANTUM INFORMATION?

Combining physics and philosophy, this is a uniquely interdisciplinary examination of quantum information science that provides an up-to-date examination of developments in this field.

The authors offer coherent definitions and theories of information, taking clearly defined approaches to considering information in connection with quantum mechanics, probability, and correlations. Concepts addressed include entanglement of quantum states, the relation of quantum correlations to quantum information, and the meaning of the informational approach for the foundations of quantum mechanics. Furthermore, the mathematical concept of information in the communicational context and the notion of pragmatic information are considered.

Suitable as both a discussion of the conceptual and philosophical problems of this field and a comprehensive stand-alone introduction, this book will benefit both experienced and new researchers in quantum information and the philosophy of physics.

OLIMPIA LOMBARDI is the director of a research group in the philosophy of physics and the philosophy of chemistry. She has reviewed for several of the most prestigious journals specialized in physics and the philosophy of science, and has been awarded grants from the Foundational Questions Institute and the John Templeton Foundation.

SEBASTIAN FORTIN is a Fellow Researcher of the National Scientific and Technical Research Council in Argentina, and has spoken and published widely on quantum information theory, the foundations of quantum mechanics, and the foundations of quantum chemistry.

FEDERICO HOLIK is a Fellow Researcher of the National Scientific and Technical Research Council in Argentina. His research focuses on the foundations of quantum mechanics, quantum information theory, the interpretation of quantum probabilities, and the study of the logical, algebraic, and geometrical aspects of quantum formalism.

CRISTIAN LÓPEZ is a Visiting Fellow of the Münich Center for Mathematical Philosophy and holds a graduate research fellowship at the National Scientific and Technical Research Council in Argentina. His research focuses on the philosophy of physics and the philosophy of time.

WHAT IS QUANTUM INFORMATION?

Edited by

OLIMPIA LOMBARDI

University of Buenos Aires, Argentina, National Council of Scientific and Technical Research

SEBASTIAN FORTIN

University of Buenos Aires, Argentina, National Council of Scientific and Technical Research

FEDERICO HOLIK

National University of La Plata, Argentina, National Council of Scientific and Technical Research

CRISTIAN LÓPEZ

University of Buenos Aires, Argentina, National Council of Scientific and Technical Research

CAMBRIDGE
UNIVERSITY PRESS

University Printing House, Cambridge CB2 8BS, United Kingdom

One Liberty Plaza, 20th Floor, New York, NY 10006, USA

477 Williamstown Road, Port Melbourne, VIC 3207, Australia

4843/24, 2nd Floor, Ansari Road, Daryaganj, Delhi – 110002, India

79 Anson Road, #06–04/06, Singapore 079906

Cambridge University Press is part of the University of Cambridge.

It furthers the University's mission by disseminating knowledge in the pursuit of education, learning, and research at the highest international levels of excellence.

www.cambridge.org
Information on this title: www.cambridge.org/9781107142114
DOI: 10.1017/9781316494233

First published 2017

Printed in the United States of America by Sheridan Books, Inc.

A catalogue record for this publication is available from the British Library.

Library of Congress Cataloging-in-Publication Data
Names: Lombardi, Olimpia, editor. | Fortin, Sebastian, 1979– editor. | Holik,
Federico, editor. | López, Cristian, editor.
Title: What is quantum information? / edited by Olimpia Lombardi (Universidad de Buenos Aires,
Argentina), Sebastian Fortin (Universidad de Buenos Aires, Argentina), Federico Holik (Universidad
Nacional de La Plata), Cristian López (Universidad de Buenos Aires, Argentina).
Description: Cambridge, United Kingdom ; New York, NY : Cambridge University Press, 2017. |
Includes bibliographical references.
Identifiers: LCCN 2016057954| ISBN 9781107142114 | ISBN 1107142113
Subjects: LCSH: Physics – Philosophy. | Information theory. | Quantum theory.
Classification: LCC QC6 .W56 2017 | DDC 530.1201/154–dc23
LC record available at https://lccn.loc.gov/2016057954

ISBN 978-1-107-14211-4 Hardback

Contents

Contributors

Olimpia Lombardi University of Buenos Aires, Argentina, National Council of Scientific and Technical Research

Sebastian Fortin University of Buenos Aires, Argentina, National Council of Scientific and Technical Research

Federico Holik National University of La Plata, Argentina, National Council of Scientific and Technical Research

Cristian López University of Buenos Aires, Argentina, National Council of Scientific and Technical Research

Armond Duwell University of Montana, United States of America

Jeffrey Bub University of Maryland, United States of America

Dennis Dieks University of Utrecht, Netherlands

Juan Roederer University of Alaska Fairbanks, United States of America

Adán Cabello University of Seville, Spain

Amit Hagar University of Indiana, United States of America

David Wallace University of Southern California, United States of America

Gustavo Martín Bosyk National University of La Plata, Argentina

Guido Bellomo National University of La Plata, Argentina

Ángel Luis Plastino National University of La Plata, Argentina

Ángel Ricardo Plastino National University of La Plata, Argentina

Preface

It is not easy to work on philosophy of physics when living in the southernmost country of South America. Besides the language difficulties and the lack of a tradition on the matter, distance is the main obstacle. Much time and money is necessary to attend the best academic meetings and to visit specialized scholars: everything happens in the Northern Hemisphere.

For this reason, the Large Grant that we received from the Foundational Questions Institute (FQXi) was a kind of oasis for our academic development. It allowed us not only to attend several relevant conferences where we could present our work on information, but also made it possible to organize in Buenos Aires the international workshop "What Is Quantum Information?" that was the root from which the present book finally emerged. Our first acknowledgement, thus, goes to the FQXi. But since institutions do not exist without the people who embody them, our acknowledgement is particularly directed to Max Tegmark and Anthony Aguirre, Scientific Director and Associate Scientific Director of the FQXi, respectively, and to Kavita Rajanna, managing director, and Brendan Foster, scientific programmes consultant, for their continued support during two years of work.

The success of the meeting, nevertheless, was the result of further essential factors. The first of them was the fact that some of the most renowned international specialists on quantum information kindly accepted our invitation to participate in the workshop: Jeffrey Bub, Adán Cabello, Dennis Dieks, Armond Duwell, Christopher Fuchs, Angelo Plastino, Robert Spekkens, and Christopher Timpson. Their interesting talks and the lively final discussion were an invaluable contribution to the reflection on the concept of quantum information. Our special acknowledgement to all of them.

We also want to heartily acknowledge Gloria Dubner, Director of the Instituto de Astronomía y Física del Espacio (IAFE), who kindly offered the institute as the venue for the meeting. But the success of the workshop would not be possible without the essential assistance of the members of the Group of Philosophy of Science led by Olimpia and based in the faculties of sciences and of philosophy of the University of Buenos Aires: Hernán Accorinti, Guido Bellomo, Martín Bosyk, Mariana Córdoba, María José Ferreira Ruiz, Paula Lipko, Marcelo Losada, Juan Camilo Martínez González, and Erick Rubio. Their strong commitment and unlimited enthusiasm made the organization of the meeting an enjoyable task.

Although the workshop was the basis for the present book, we want to especially thank Amit Hagar, Juan Roederer, and David Wallace, who nicely accepted our invitation to contribute to the ongoing project by sending us their papers, which supply refreshing perspectives on the subject matter. Last, but not least, we want to express our gratitude to Cambridge University Press, in the persons of Simon Capelin, Editorial Director (physical sciences), and Philippa Cole, Editorial Assistant (physics), for their support and assistance during all the stages of this project.

Introduction

By analogy with the Industrial Revolution, at present many people talk about an Information Revolution that began in the mid-twentieth century and continues to this day. Although triggered by the advent of digital technologies and established by the proliferation of digital computers, the drastic changes rapidly exceeded the limits of technology to pervade all aspects of social life. Nowadays, information shapes all our everyday activities and thoughts.

Given this situation, it is not surprising that, during the past decades, philosophy has begun to focus its attention on the search of an elucidation of the notion of information. The many dimensions of information make this task particularly interesting from a philosophical viewpoint, but, at the same time, attempt against a unified answer to the problem. At present, different interpretations of the notion of information coexist, sometimes as the consequence of implicitly conflating its different meanings, but in many cases also as the result of the multiple facets of the concept.

At the same time, new interpretive problems have arisen with the advent of the research field called "quantum information theory." Those problems combine the difficulties in the understanding of the concept of information with the well-known foundational puzzles derived from quantum mechanics itself. Of course, interpretive issues were not an obstacle to the huge development of quantum information theory as a scientific area of research, where new formal results multiply rapidly. Nevertheless, the question "What is quantum information?" is still far from having an answer on which the whole quantum information community agrees.

It is in this context that the question about the nature of quantum information deserves to be considered from a conceptual viewpoint. The aim of this volume is, precisely, to address the issue from several and varied perspectives, which makes manifest its different aspects and its many implications. With this purpose, the chapters of this volume are organized in three parts. Part I, "The Concept of Information," groups the chapters mainly devoted to inquiring into the concept itself and its relationships with other notions, such as knowledge, representation, and manipulation. In Part II, "Information and Quantum Mechanics," the links between informational and quantum issues enter the stage. Finally, Part III, "Probability, Correlations, and Information," addresses the subject matter by considering how the notions of probability and correlation underlie the concept of information in different problem domains.

Part I opens with the chapter "About the Concept of Information," where Sebastian Fortin and Olimpia Lombardi begin by introducing some relevant distinctions that allow them to focus on mathematical information in the communicational context. In this context, after discussing the definition of some magnitudes involved in the Shannon formalism, the chapter deals entirely with interpretive matters. First, three interpretations of the concept of information are introduced, stressing their differences and specific difficulties. Then, the question about the existence of two qualitatively different kinds of information, classical and quantum, is addressed. On the basis of the previous discussion, the authors advocate for a theoretically neutral interpretation of information.

The main aim of the second chapter of this first part, "Representation, Interpretation, and Theories of Information," by Armond Duwell, is to stress the vital importance of representational and interpretive aspects for understanding the definition of information provided by Christopher Timpson. With this purpose, the chapter begins with discussing some basic features of representation and interpretation of theories. Then, two potential problems of Timpson's definition in Shannon information theory and in quantum information theory, respectively, are considered; the argumentation is directed to show that the resolution to those problems depends on recognizing how important users of the information theories are in determining what constitutes successful quantum information transfer. On this basis, the author concludes that Timpson's definition of information functions perfectly well and correctly elucidates what information is; moreover, specializations of this definition to various theories illustrate the differences in different types of information.

The third and final chapter of the opening part, "Information, Communication, and Manipulability," by Olimpia Lombardi and Cristian López, aims at supplying adequate criteria to identify information in a communicational context. For this purpose, the chapter begins by considering the different interpretations of Shannon's formalism that can be implicitly or explicitly found in the literature, and the additional challenges raised by the advent of entanglement-assisted communication. This analysis shows that the communication of information is a process that involves a certain idea of causation and the asymmetry implicit in it. On this basis, the authors claim that the manipulability accounts of causation supply the philosophical tools to characterize the transmission of information in a communicational context, and that many conundrums around the concept of information in this context are solved or simply vanish in the light of a manipulability view of information.

Part II begins with Jeffrey Bub's chapter, "Quantum versus Classical Information." Bub opens his chapter by stressing that the question "What is quantum information?" has two parts: first, "what is information?" second, "what is the difference between quantum information and classical information?" and he proposes an answer to the second question. With this purpose, Bub begins by supplying a characterization of intrinsic randomness, and then shows that the nonlocal correlations of entangled quantum systems are only possible if the measurement outcomes on the separate systems are intrinsically random events. Then, intrinsic randomness increases the possibilities for information processing, essentially because new sorts of correlations are possible that cannot occur in a classic world. On this basis, the author concludes that intrinsic randomness marks the difference between quantum and classical information: quantum information is a type of information that is only possible in a world in which there are intrinsically random events.

In the following chapter of Part II, "Quantum Information and Locality," Dennis Dieks begins by recalling that the surprising aspects of quantum information arc due to two distinctly non-classical features of the quantum world: first, different quantum states need not be orthogonal, and, second, quantum states may be entangled. He focuses on the concept of entanglement, since it leads, via non-locality, to those forms of communication that go beyond what is classically possible. In particular, he analyzes the significance of entanglement for the basic physical concepts of "particle" and of "localized physical system." According to the author, in general the structure of quantum mechanics is at odds with an interpretation in terms of particles, which may be localized. This leads him to the conclusion that quantum mechanics is best seen as not belonging to the category of space-time theories: the resulting picture of the quantum world is relevant for understanding in what sense quantum theory is non-local, and this in turn sheds light on the novel aspects of quantum information.

In his chapter, "Pragmatic Information in Quantum Mechanics," Juan Roederer argues that information is essentially a pragmatic notion. The chapter begins by distinguishing between two categories of interactions between bodies or systems: force-driven, which operates in the entire spatial-temporal domain, and information-driven, which leads to the definition of information as a pragmatic concept. Pragmatic information is defined as that which represents a physical, causal, and univocal correspondence between a pattern and a specific macroscopic change mediated by some complex interaction mechanism. On the basis of this definition, pragmatic information in itself does not operate in the quantum domain. According to the author, to the extent that information is pragmatic, talking about inaccessible or hidden information in quantum states makes no sense: quantum mechanics can only provide real – pragmatic – information by means of natural or deliberate macroscopic imprints left by a composite quantum system, which as a single whole interacts irreversibly with the surrounding macroscopic world.

In the last chapter of Part II, "Interpretations of Quantum Theory: A Map of Madness," Adán Cabello stresses the fact that, at present, physicists do not yet agree about what quantum theory is about, and argues that it is urgent to solve this problem. In order to contribute to the solution, he classifies the interpretations of quantum theory into two types, according to whether the probabilities of measurement outcomes are determined by intrinsic properties of the observed system or not. Cabello considers that these two types of interpretations are so radically different that there must be experiments that, when analyzed outside the framework of quantum theory, lead to different empirically testable predictions.

Part III, devoted to probabilities and correlations, begins with the chapter "On the Tension between Ontology and Epistemology in Quantum Probabilities," where Amit Hagar proposes a physical underpinning of quantum probabilities, which is dynamical, finitist, operational, and objective. According to this operationalist view, which dispels the metaphysics that surrounds the quantum state, finite-resolution measurement outcomes are taken as primitive and basic building blocks of the theory, and quantum probabilities are objective dynamical transition probabilities between finite-resolution measurement results. As a consequence, nonrelativistic quantum mechanics is seen as a phenomenological, "effective" theory, whose mathematical structure – the Hilbert space – rather than a fundamental structure that requires interpretation, is a tool for computing the probabilities of future states of an underlying deterministic and discrete process, from the inherently and objectively limited knowledge we have about it. According to the author, this view of probabilities can qualify as an objective alternative to the subjective view that the quantum information theoretic approach adheres to.

In his chapter, "Inferential versus Dynamical Conceptions of Physics," David Wallace addresses the issue of probabilities in physics by contrasting two possible attitudes towards a given branch of physics: inferential, as concerned with an agent's ability to make predictions given finite information, and dynamical, as concerned with the dynamical equations governing particular degrees of freedom. He contrasts these attitudes in classical statistical mechanics, in quantum mechanics, and in quantum statistical mechanics. In this last case, he argues that the quantum-mechanical and statistical-mechanical aspects of the question become inseparable. On this basis, the conclusion of the chapter is that the particular attitude adopted – whether to conceive of a given field in physics as a form of inference or as a study of dynamics – plays a central role in the foundations of quantum theory, and the exact same role in the foundations of statistical mechanics once it is understood quantum mechanically.

The chapter "Classical Models for Quantum Information," by Federico Holik and Gustavo Martín Bosyk, faces the question about the ontological status of quantum information. The first part of the chapter emphasizes the existence of probabilistic models that go beyond the classical and quantal realms, and the possibility of performing informational protocols in those models. On this basis, the authors argue that a generalized information theory can be conceived. In the second part, the question about the ontological reference of those probabilistic models is addressed. For this purpose, the authors recall the existence of many examples of physical systems built by means of an essentially classical ontology, but that are modeled by formal structures with quantum features. Their significance relies on the fact that they can be used to perform quantum information protocols. This fact points to the need of exploring the ontological implications of those simulations for the concept of quantum information.

The last chapter of Part III, "On the Relative Charecter of Quantum Corrlations," by Ángel Luis Plastino, Guido Bellomo, and Ángel Ricardo Plastino, revisits the concepts of entanglement and discord from a generalized perspective that focuses on the relational aspect of the term "correlation" with respect to the states and observables involved. From the fact that the concept of correlation is inherently relative to the non-unique partition of a system into subsystems, the authors favor a description-dependent view of the quantum correlations that provides what they call a second stage of relativity. On this basis, they propose generalized definitions for entanglement and discord. Moreover, the authors argue that the relative character of quantum correlations imposes restrictions on the classical appearance of the quantum. In particular, they prove that some types of quantum correlations may appear as classical correlations when certain relevant observables in a larger Hilbert space are measured. Therefore, the classical appearance of the quantum world is also relative to a given description.

As this brief overview shows, the question about what quantum information is can be addressed from many different perspectives, some of them complementary, others conflicting with each other. This plurality precisely reveals to what extent the community of quantum physicists and philosophers of physics is far from a consensus about the answer of that question. This volume is intended as a contribution to the discussion about the conceptual foundations of an exciting field like quantum information theory.

Part I

The Concept of Information

1

About the Concept of Information

SEBASTIAN FORTIN AND OLIMPIA LOMBARDI

1 Introduction

As with other concepts that marked an era in the past, the clue of our time seems to lie in the concept of information. The appeal of the slogan "we live in the age of information" is strongly justified: we can perceive the pervasiveness of information in our lives as embodied in the explosion of highly advanced technologies devoted to computing and communication. But behind this materialization, at present, information has become an immaterial source of power, not only in the social domain, but also in the political and economic spheres. It is, then, not surprising that science has not been left out of this "information revolution": nowadays the concept of information has permeated almost all scientific disciplines, from physics and chemistry to biology and psychology.

In the face of this situation, since the past decades philosophy has begun to address the question of the content and the role of the concept of information in different domains. The present chapter is framed in this general trend: the aim here is to discuss the technical concept of information in the communicational context, whose classical locus is the paper where Claude Shannon (1948) introduces a precise formalism designed to solve certain specific technological problems in communication engineering. For this purpose, this chapter is organized as follows. Section 2 will introduce relevant distinctions that will allow us to focus on the concept of information in the context of interest. In Section 3, Shannon's formalism will be presented and some technical issues about the definition of the relevant magnitudes will be discussed. Section 4 will be devoted to interpretive matters: three interpretations of the concept of information will be introduced, stressing their differences and specific difficulties. Section 5 will contend with information in the quantum domain: the question about the existence of two qualitatively different kinds of information, classical and quantum, will be addressed. Finally, on the basis of the previous discussion, Section 6 will include some final remarks about the advantages of a theoretically neutral interpretation of information.

2 Finding the Way through Bifurcations

The philosophical analysis of the concept of information has resulted in a growing amount of literature devoted to this subject matter from very different perspectives. The purpose of this section is not to offer an exhaustive map of the land in the philosophical literature on information, a task that surely exceeds the capacity of a single chapter. The aim here is to introduce some relevant distinctions that will allow us to focus on the concept of information that will be the subject of interest in the subsequent sections.

2.1 *'Information': Everyday Use versus Technical Use*

The use of the term 'information' is not confined to the language of science or of philosophy. On the contrary, the term is strongly present in everyday language, with a general meaning that, as in all cases, resulted from a long and complex historical process. Related to the idea of knowing as apprehending the *form* of objects, prevailing in antiquity and the Middle Ages, in modern times, the term 'information' disappeared from the philosophical discourse, but gained popularity in everyday language, with the link to knowledge that has survived until today (see Adriaans 2013).

An author who stresses the difference between the everyday use and the technical use of the term 'information' is Christopher Timpson (2004, 2013), who takes the analogy between "truth" and "information" as a departing point to support his claim that 'information' is an abstract noun: "Austin's aim was to de-mystify the concept of truth, and make it amenable to discussion, by pointing to the fact that 'truth' is an abstract noun. So too is 'information'" (2013: 10). Timpson recalls that very often abstract nouns arise as nominalizations of various adjectival or verbal forms. On this basis, he extends the analogy between truth and information: "Austin leads us from the substantive 'truth' to the adjective 'true.' Similarly, 'information' is to be explained in terms of the verb 'inform'" (2013: 11). In turn, "[t]o inform someone is to bring them to know something (that they did not already know)" (2013: 11). In other words, the meaning of 'information' in everyday language is given by the operation of bringing knowledge and, therefore, the word is an abstract noun that does not refer to something concrete that exists in the world.

In this chapter, we will not analyze the use of the term 'information' in ordinary language, but we are interested in the technical use of the term, which is nevertheless far from being univocal.

2.2 *The Technical Domain: Semantic and Mathematical Information*

In the technical domain, the first distinction to be introduced is that between a semantic and a non-semantic view of information.

According to the first view, information is something that carries semantic content (Bar-Hillel and Carnap 1953; Bar-Hillel 1964; Floridi 2010, 2011); it is therefore strongly related to semantic notions such as reference, meaning, and representation. In general, semantic information is carried by propositions that intend to represent states of affairs; so, it has intentionality, "aboutness," that is, it is directed to other things. And although it remains controversial whether false factual content may qualify as information (see Graham 1999; Fetzer 2004; Floridi 2004, 2005; Scarantino and Piccinini 2010), semantic information maintains strong links with the notion of truth. At present there is a well-developed field of research in the philosophy of semantic information (see, e.g., Adriaans and Van Benthem 2008, and the Web site of the Society for the Philosophy of Information), in the context of which many strongly technical views of semantic information are proposed (just to mention some of them: Dretske 1981; Barwise and Seligman 1997; Floridi 2011; for a wide, updated source of references, see Floridi 2015; for an analysis of Dretske's proposal, see Lombardi 2005).

Non-semantic information, also called 'mathematical,' is concerned with many formal properties of different kinds of systems, among them the best known are the compressibility properties of sequences of states of a system and the correlations between the states of two systems, independently of the meanings of those states. We will focus on mathematical information; however, the mention of "mathematical information" does not specify yet what we are taking about.

2.3 The Many Faces of Mathematical Information

In the domain of mathematical information, different contexts can be distinguished, each one with its particular formal resources to deal with specific goals. The two traditional contexts are the computational and the communicational.

In the *computational context*, information is something that has to be computed and stored in an efficient way. In this framework, the algorithmic or Kolmogorov complexity measures the minimum resources needed to effectively reconstruct an individual message (Solomonoff 1964; Kolmogorov 1965, 1968; Chaitin 1966): it supplies a measure of information for individual objects taken in themselves, independently of the source that produces them. In the theory of algorithmic complexity, the basic question is the ultimate compression of individual messages. The main idea that underlies the theory is that the description of some messages can be compressed considerably if they exhibit enough regularity. The Kolmogorov complexity of a message is, then, defined as the length of the shortest possible program that produces it in a Turing machine. Many information theorists, especially computer scientists, regard algorithmic complexity as more fundamental than Shannon entropy as a measure of information (Cover and Thomas 1991: 3), to the extent that algorithmic complexity assigns an asymptotic complexity to an individual message without any recourse to the notion of probability (for a discussion of the relation between Shannon entropy and Kolmogorov complexity, see Cover and Thomas 1991: chapter 7; Lombardi, Holik, and Vanni 2016b).

In the *communicational context,* whose classical formalism was formulated by Claude Shannon (Shannon 1948; Shannon and Weaver 1949), information is primarily something to be transmitted between two points for communication purposes. The formalism Shannon proposed was designed to solve certain specific technological problems in communication engineering, in particular, to optimize the transmission of information by means of physical signals whose energy and bandwidth is constrained by technological and economic limitations. Shannon's theory is purely quantitative; it ignores any issue related to informational content: "[the] semantic aspects of communication are irrelevant to the engineering problem. The significant aspect is that the actual message is one selected from a set of possible messages" (1948: 379).

Although the best known, the computational and the communicational contexts are not the only mathematical contexts in which the concept of information is present. For instance, in an *inferential context,* the interest is to find a universally good prediction procedure on the basis of the possessed data, where "good" usually involves a version of Occam's Razor: "The simplest explanation is best." The *thermodynamic context* is devoted to relate information and entropy and to explain the entropy increase in terms of informational concepts and arguments. In a *gambling context,* the problem is to use informational resources to formalize a gambling game, by representing the wealth at the end of the game as a random variable and the gambler as a subject that tries to maximize that variable. These are only some of the non-classical contexts that show that, at present, information theory is a very wide body of formal knowledge with many different interests and applications (see Lombardi, Fortin, and Vanni 2015). In this chapter, we are interested in the communicational context, where the concept of information acquires a specific formal treatment, given by Shannon's theory.

3 The Formalism in the Communicational Context: Shannon's Theory

According to Shannon (1948; see also Shannon and Weaver 1949), a general communication system consists of five parts (see Figure 1.1):

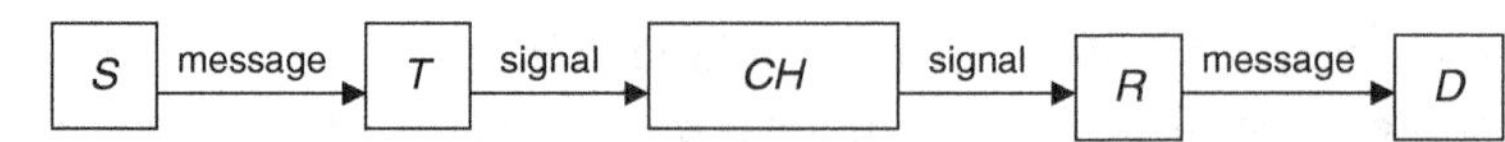

Figure 1.1. General communication system.

– A *source S*, which generates the message to be received at the destination.
– A *transmitter T*, which turns the message generated at the source into a signal to be transmitted. In the cases in which the information is encoded, encoding is also implemented by this system.
– A *channel CH*, that is, the medium used to transmit the signal from the transmitter to the receiver.
– A *receiver R*, which reconstructs the message from the signal.
– A *destination D*, which receives the message.

The source S is a system with a range of possible states $s_1, \ldots, s_n$ usually called *letters*, whose respective probabilities of occurrence are $p(s_1), \ldots, p(s_n)$ (the definitions can be extended to the continuous case; see, e.g., Cover and Thomas 1991). Since S produces sequences of states, usually called *messages*, the *entropy of the source S* is computed as:

$$H(S) = \sum_{i=1}^{n} p(s_i) \log\left(1/p(s_i)\right) \tag{1}$$

Analogously, the destination D is a system with a range of possible states $d_1, \ldots, d_m$, with respective probabilities $p(d_1), \ldots, p(d_m)$, and the *entropy of the destination D* is computed as:

$$H(D) = \sum_{j=1}^{m} p(d_j) \log\left(1/p(d_j)\right) \tag{2}$$

$H(S)$ and $H(D)$ are related through the *mutual information* $H(S; D)$, which represent the amount of information produced at the source S and recovered at the destination D:

$$H(S; D) = H(S) - E = H(D) - N \tag{3}$$

where the *equivocity E* represents the amount of information generated at S but not received at D, and the *noise N* represents the amount of information received at D but not generated at S. E and N are measures of the dependence between source and destination and, therefore, are functions not only of S and D, but also of the conditional probabilities $p(d_j/s_i)$ of the occurrence of the state d_j at D given the occurrence of the state s_i at S. In turn, the *channel capacity C* is defined as:

$$C = \max_{p(s_i)} H(S; D) \tag{4}$$

where the maximum is taken over all the possible distributions $p(s_i)$ at the source. C measures the largest amount of information that can be *transmitted* over the communication channel CH.

In the transmitter T, coding is a mapping from the alphabet $A_S = \{s_1, \ldots, s_n\}$ of letters of the source S to the set of finite length strings of *symbols* from the code alphabet $A_C = \{c_1, \ldots, c_q\}$ of T. In general, those strings, called *code-words*, do not have the same length: the code-word w_i, corresponding to the letter s_i, has a length l_i. This means that coding is a fixed-to-variable-length mapping. Therefore, the *average code-word length L* can be defined as:

$$L = \sum_{i=1}^{n} p(s_i) \; l_i \tag{5}$$

L indicates the compactness of the code: the lower the value of L, the more efficient the coding, that is, the fewer resources needed to encode the messages. The *noiseless coding theorem* (or *First Shannon Theorem*) proves that, for very long messages (strictly speaking, for messages of length $N \to \infty$), there is an optimal encoding process such that the average code-word length L is as close as desired to the lower bound $L_{\min}$ for L:

$$L_{\min} = \frac{H(S)}{\log q} \tag{6}$$

where, when $H(S)$ is expressed in bits, log is the logarithm to base 2. When $H(S)$ is expressed in bits and the code alphabet has two symbols (an alphabet of binary digits, $q = 2$), then $\log_2 q = \log_2 2 = 1$, and the noiseless coding theorem establishes the direct relation between the entropy of the source and the lower bound $L_{\min}$ of the average code-word length L.

In turn, the *noisy coding theorem* (or *Second Shannon Theorem*) proves that the information transmitted over a communication channel can be increased without increasing the probability of error as long as the communication rate is maintained below the channel capacity. In other words, the channel capacity is equal to the maximum rate at which the information can be sent over the channel and recovered at the destination with a vanishingly low probability of error.

Although clear and extensively applied, Shannon's theory still admits certain disagreements about the definition of its formal concepts, which will be considered in the following subsections.

3.1 Units of Measurement

In the calculation of the magnitudes representing information, the choice of a logarithmic base amounts to a choice of a unit of measurement for information. If the base 2 is used, the resulting unit is called *'bit.'* But the natural logarithm can also be used, and in this case the unit of measurement is the *nat,* contraction of "natural unit." And when the logarithm to base 10 is used, the unit is the *Hartley.* Although trivial in the stage of generating information, the possibility of using different units to measure information is relevant when coding is taken into account. In fact, as explained previously, the noiseless coding theorem shows that the entropy of the source $H(S)$ is proportional to L_{min} (see eq. (6)): the constant of proportionality depends on which unities are used to express the entropy of the source and how many symbols the code alphabet has, and these two aspects are completely independent. The possibility of using different units to quantify information shows the difference between the amount of information generated by the source and the number of symbols necessary to encode that information.

On the other hand, for a long time it was quite clear for communication engineers that "bit" was a unit of measurement, and that the fact that a different unit can be used did not affect the very nature of information. However, with the advent of quantum information, the new concept of *"qubit"* entered the field: a qubit is primarily conceived not as a unit of measurement of quantum information, but as a quantum system of two-states used to encode the information of a source. This way of talking about qubits has gradually seeped into Shannon's theory in the talk about bits. This process led to a progressive reification of the concept of bit, which now is also – and many times primarily – conceived as referring to a classical system of two states. The conceptual problem is not the present-day ambiguity of the term 'bit,' but the frequent conflation of the two meanings of the term: "The Shannon information $H(X)$ measures in bits (classical two-state systems) the resources required to transmit all the messages that the source produces" (Timpson 2006: 592). This undifferentiated use of the term 'bit' is conceptually unsatisfactory: it is like confusing a meter with the Prototype Meter bar, an object made of an alloy of platinum and iridium and stored in the Bureau International des Poids et Mesures in Sèvres; and saying that the Shannon information $H(X)$ gives a measure "in bits (classical two-state systems)" is like saying that the length L gives a measure "in meters (platinum-iridium bars)."

In order to avoid this kind of confusion about the concept of bit, it might be appropriate to follow the suggestion of Carlton Caves and Christopher Fuchs (1996), who propose to use the term *'cbit'* to name a two-state classical system used to encode Shannon information, by analogy with the two-state quantum system, the qubit, used to encode quantum information (or, at least to encode information by means of quantum resources). This terminology keeps explicit the distinction between the entropy of the source, which is usually measured in bits but can be measured in other units, the alphabet by means of which the messages of the source are encoded, which consist of a number q

of symbols, and the systems of q states used to physically implement the code alphabet.

3.2 Amounts or Average Amounts?

In some presentations of the theory, the entropies $H(S)$ and $H(D)$ are computed directly in terms of the probabilities of the states of the source and the destination. However, since eqs. (1) and (2) have the form of weighted averages, $H(S)$ and $H(D)$ are usually defined as *average amounts of information per letter* generated by the source and received by the destination, respectively (see, e.g., Lombardi 2004; Bub 2007):

$$H(S) = \sum_{i=1}^{n} p(s_i)\, I(s_i) \tag{7}$$

$$H(D) = \sum_{j=1}^{m} p(d_j)\, I(d_j) \tag{8}$$

where $I(s_i) = \log\left(1/p(s_i)\right)$ and $I(d_j) = \log\left(1/p(d_j)\right)$ are the individual amounts of information corresponding to the occurrence of the single state s_i at the source and of the single state d_j at the destination, respectively. From this usual perspective, a single letter of the source is a particular kind of message and conveys information. For instance, Ralph Hartley – whose work Shannon explicitly acknowledged as one of the bases of his proposal – states that, in the case of equiprobable alternatives, "[t]he information associated with a single selection is the logarithm of the number of symbols available" (1928: 541).

By contrast to this usual presentation of Shannon's theory, for Christopher Timpson, it makes no sense to talk about individual amounts of information: the entropies $H(S)$ and $H(D)$ are computed in terms of the probabilities of the source and the destination, but they are not defined in terms of a more basic form of information. From this perspective, individual messages do not convey information, and the entropy $H(S)$ is a property of the source of information S: "It is essential to realize that 'information' as a quantity in Shannon's theory is not associated with individual messages, but rather characterizes the source of the messages" (Timpson 2013: 21). As a consequence, $H(S)$ and $H(D)$ cannot be conceived as *average* amounts, since only in terms of individual magnitudes averages can be significantly computed as such.

The distinction between conceiving the entropies of the source and the destination as amounts of information or average amounts of information might seem an irrelevant detail. However, this is not the case when we are interested in elucidating the very notion of information – in Shannon's sense – since, as it will be explained in the next subsection, assuming the conceptual priority of the entropies over the individual amounts of information makes it possible to define the concept of information in terms of the coding theorems.

3.3 Generation and Coding

A simple and usual reading of Shannon's theory is to consider that the source S generates information, which is quantified – absolutely or in average – by $H(S)$ and measured in some unit of measurement, in general, in bits. This information, carried by the messages produced at the source, is encoded in the transmitter by means of symbols embodied in physical systems: the output of the transmitter is a signal that conveys the encoded information. From this viewpoint, information can be defined and quantified at the stage of the source, and the First Shannon coding theorem shows that the amount of information generated at the source is proportional to the average code-word length in *optimal* coding.

However, another reading has been proposed: information is defined in terms of the coding theorems: "the coding theorems that introduced the classical (Shannon, 1948) and quantum (Schumacher, 1995) concepts of information$_t$ [the technical concept of information] do not merely define measures of these quantities. They also introduce the concept of *what it is* that is transmitted, *what it is* that is measured" (Timpson 2008: 23, emphasis in the original). From this perspective, the meaning of the entropy $H(S)$ is defined by the First Shannon theorem as the minimum amount of channel resources required to encode the output of the source in such a way that any message produced may be accurately reproduced at the destination (Timpson 2008: 27; 2013: 37, 43). In other words, asking how much information a source produces is asking to what degree the output of the source is compressible. This view is in resonance with depriving individual messages of the property of carrying information: $H(S)$ is not an average of individual amounts of information previously defined, but is defined directly as compressibility. Moreover, this position turns the result of a theorem into a definition: now the entropy $H(S)$ is not defined by eq. (1) but by eq. (6), and eq. (1) must be obtained as the result of a mathematical proof.

A first point to notice is that only when $H(S)$ is expressed in a unit of measure defined by $\log_n$ and the code alphabet has n symbols, the noiseless coding theorem identifies the entropy of the source and the lower bound $L_{\min}$ of the average code-word length. In the general case, the entropy of the source is only proportional to $L_{\min}$ (see eq. (6)), and the constant of proportionality depends on the unities used to express the entropy of the source and the number of symbols of the code alphabet, and these two aspects are completely independent. And since the code alphabet belongs to the transmitter, when information is defined as compressibility, a property of the source (the compressibility of the messages it produces) does not depend only on its own features, but also on a feature of a different system.

On the other hand, if the noiseless coding theorem says *what* information is, $H(D)$, E and N do not represent information since they are not involved in the theorem. But, then, it is not clear how they can participate with $H(S)$ in algebraic operations as in eq. (3). Moreover, since the result of the noiseless coding theorem corresponds to ideal coding, it is not clear how to talk of information in the case of non-ideal coding: can we still say that the same amount of information can be better or worse encoded? In turn, since the coding theorem is proved in the case of very long messages, strictly speaking, for messages of length $N \rightarrow \infty$, one wonders whether short messages can be conceived as embodying information to the extent that they are not covered by the theorem.

When explaining the elements of the general communication system, Shannon (1948) characterizes the transmitter as a system that operates on the message coming from the source in some way to produce a signal suitable for transmission over the channel. He also stresses that, in many cases, such as in telegraphy, the transmitter is also responsible for encoding the source messages. However, in certain cases the message is not encoded; for instance, in traditional telephony the transmitter operates as a mere transducer, by changing sound pressure into a proportional electrical current. If information is defined in terms of the noiseless coding theorem, it is not easy to talk about information when no coding is involved.

The definition of information as compressibility does not establish a relevant distinction between the stage of generation of information and the stage of coding this information, and this leads to the consequences pointed out earlier (for a detailed discussion, see Lombardi, Holik, and Vanni 2016b). It is clear that none of them is an insurmountable objection: each difficulty can be answered by a particular argument. But the traditional view removes all the difficulties in a single move: the – average – amount of information produced by the source is defined by the features of the source itself (the probabilities of its states), and is independent of coding – even of the very fact that the messages are encoded or not – when that information is later encoded, the coding noiseless theorem says how this quantity of information previously defined can be ideally encoded.

3.4 *The Success of Communication*

As Shannon stresses, in communication, "[t]he significant aspect is that the actual message is one *selected from a set* of possible messages" (1948: 379, emphasis in the original). In fact, the states d_j of the destination system D can be any kind of states, completely different from the states s_i of the source system S: the states of the source and the states of the destination may be of a completely different nature: for instance, the source may be a dice and the destination a dash of lights; or the source may be a device that produces words in English and the destination a device that operates a machine. In any case, the goal of communication is to identify at the destination which sequence of states was produced by the source, and communication is successful when the goal is met.

This characterization of the goal of communication might lead us to suppose that the criterion for success is given by a one-to-one mapping from the set of the letters of the source to the set of the letters of the destination. Armond Duwell is clear in this sense:

The most natural way to think about success internal to the Shannon theory is to have a one-[to]-one mapping from the set of symbols that characterize the source to the set of symbols that characterize the destination. This mapping establishes an identity between the symbols that characterize the source and destination that allows one to determine when communication is successful, i.e. when the "same" sequence of symbols is reproduced at the destination that was produced by the source.

(2008: 200)

From a similar perspective, Timpson characterizes the goal of communication as reproducing at the destination another token of the same type as that produced at the source (Timpson 2008: 25; 2013: 23). But the type-token distinction is *generalized* in terms of *sameness of pattern or structure*: "the distinction may be generalized. The basic idea is of a pattern or structure: something which can be repeatedly realized in different instances" (Timpson 2013: 18; for a criticism of this view of the type-token distinction, see Lombardi, Fortin, and López 2016).

This view of characterizing the success of communication, although seemingly natural, forgets the possibility of noisy situations. In fact, only in the so-called "deterministic case" are the set of the source states and the set of the destination states linked by a one-to-one function. But in a noisy case with no equivocity, the mapping is one-to-many (see, e.g., Cover and Thomas 1991: 184–185): a single state of the source, say s_1, may lead to different states of the destination, say d_1 and d_2 with non-zero conditional probabilities $p(d_1/s_1)$ and $p(d_2/s_1)$. Nevertheless, the state of the source can be identified by means of the state of the destination. As Thomas Cover and Joy Thomas explain in their widely used book of information theory, "What do we mean when we say that A communicates with B? We mean that the physical acts of A have induced a desired physical state in B"; the communication is successful when A and B "agree on what was sent" (1991: 183). In this technical context, noisy situations cannot be disregarded, since they are the cases of real interest in the practice of communication engineering.

4 Interpreting Information in the Communicational Context

Shannon's original article was immediately followed by many works of application to fields such as radio, television, and telephony. At present, Shannon's theory has become a basic element of communication engineers' training. These well-known facts might lead us to suppose that there are no disagreements about what information is. However, this is not the case: the concept of information is still a focus of much debate, even in the technical domain.

4.1 The Epistemic Interpretation

The epistemic interpretation of information agrees with the everyday view in endowing knowledge with a central role: information is something that provides knowledge, something that modifies the state of knowledge of those who receive it. This agreement may suggest that the link between information and knowledge is a feature of the everyday notion of information and not of the technical concepts (see Timpson 2004, 2013). However, the literature on the subject shows that that link can frequently be found both in philosophy and in science.

For instance, in the field of philosophy, Fred Dretske says: "information is a commodity that, given the right recipient, is capable of yielding knowledge" (1981: 47). Some authors devoted to special sciences are also persuaded that the core meaning of the concept of information, even in its technical sense, is linked to the concept of knowledge; in this trend, Jon M. Dunn defines information as "what is left of knowledge when one takes away believe, justification and truth" (2001: 423). Also physicists frequently speak about what we know or may know when dealing with information. For instance, Anton Zeilinger even equates information and knowledge when he says that "[w]e have knowledge, i.e., information, of an object only through observation" (1999: 633) or, with Časlav Brukner, "[f]or convenience we will use here not a measure of information or knowledge, but rather its opposite, a measure of uncertainty or entropy" (2009: 681–682). In a traditional textbook about Shannon's theory applied to engineering, one can also read that information "is measured as a difference between the state of knowledge of the recipient before and after the communication of information" (Bell 1957: 7). Although not regarding Shannon's theory but in the quantum context, Christopher Fuchs adheres to Bayesianism regarding probabilities and, as a consequence, advocates for an epistemic interpretation of information (see Caves, Fuchs, and Schack 2002).

Even though from the epistemic perspective information is not a physical item, in general it is assumed that the possibility of acquiring knowledge about the source of information by reading the state of the destination is rooted in the nomic connection between source and destination, that is, in the lawfulness of the regularities underlying the whole situation. In particular, the conditional probabilities used to compute the relevant magnitudes of the theory must be determined by natural laws, which directly or indirectly establish the links between source and destination. In fact, if those conditional probabilities represented accidental, merely de facto correlations, the states at the destination would tell us nothing about what happened at the source.

In spite of its intuitiveness, the epistemic interpretation is not free of conceptual difficulties. Whereas the mutual information $H(S;D)$ can be easily interpreted as a measure of the knowledge about the source obtained at the destination, noise and equivocation do not measure knowledge, but, on the contrary, are obstacles to knowledge acquisition. It is not easy to see how noise, which can be produced outside of the communication arrangement and has no relation with the source of information (think, e.g., white noise in a radio receiver), can be conceived as something carrying or yielding knowledge. A way out of this problem might be to suppose that only the entropies of source and destination and the mutual information, but not noise and equivocation, can be meaningfully conceptualized as measures of knowledge. But this answer would lead to admitting the possibility of adding and subtracting variables referring to different kinds of items, in this case knowledge and something different from knowledge (see, e.g., eq. (3)), a practice absolutely not allowed in mathematized sciences.

Another problem of the epistemic view concerns the relationship between information and communication. Let us consider a TV transmitter T that broadcasts a signal received by two TV sets A and B: although there is no physical interaction between the two TV sets, the correlations between their states are not accidental, but result from the physical dependence of those states on the states of T. Therefore, from an epistemic interpretation, nothing prevents us from admitting the existence of an informational link between the two TV sets: we can define a communication channel between A and B because it is possible to learn something about B by looking at A and vice versa:

from a theoretical point of view [. . .] the communication channel may be thought of as simply the set of depending relations between [a system] S and [a system] R. If the statistical relations defining equivocation and noise between S and R are appropriate, then there is a channel between these two points, and information passes between them, even if there is no direct physical link joining S with R.

(Dretske 1981: 38)

Although consistent, something in this situation sounds odd when one considers that information is related to communication. In fact, communication implies that, at some place, somebody does something that has consequences somewhere else.

But in the case of the two TV sets, nothing can be done, say, at the *A* end that will affect what happens in the *B* end. In other words, the change of the state of *A* cannot be used to control the state of *B*; so, something of the usual conception of communication is missing.

These difficulties of the epistemic interpretation of information in the technical communication context do not imply that the interpretation is completely wrong. On the contrary, it may be fruitfully applied in psychology and in cognitive sciences, where the concept of information can be used to conceptualize the human abilities of acquiring knowledge (see, e.g., Hoel, Albantakis, and Tononi 2013). The epistemic interpretation might also serve as a basis for the philosophically motivated attempts to add a semantic dimension to a formal theory of information (MacKay 1969; Nauta 1972; Dretske 1981).

4.2 The Physical Interpretation

According to the physical interpretation, information is a physical magnitude: "Information is physical" (Landauer 1991: 23). This is the position of many physicists and most engineers, for whom the link with knowledge is not a central issue, since the transmission of information can be used only for control purposes, such as operating a device at the destination end by modifying the state of the source. In this interpretive context, information is usually compared with energy, which entered the domain of physics as a mere tool to describe what we can do with physical systems – to perform work – but gradually became an essential item that plays a central unifying role in physics: energy is an item essentially present in absolutely all contemporary physical theories. In the light of the strong presence of the concept of information in present-day physics, several authors (Stonier 1990, 1996; Rovelli, personal communication) consider that it is following a historical trajectory analogous to that followed by the concept of energy in the nineteenth century. On the basis of this general characterization, there are nevertheless different ways of advocating the physical interpretation.

The *strong* physical interpretation is linked with the idea expressed by the well-known dictum "no information without representation": the transmission of information between two points of the physical space necessarily requires an information-bearing signal, that is, a physical process propagating from one point to the other. Rolf Landauer is an explicit defender of this position when he claims that

[i]nformation is not a disembodied abstract entity; it is always tied to a physical representation. It is represented by engraving on a stone tablet, a spin, a charge, a hole in a punched card, a mark on a paper, or some other equivalent.

(1996: 188; see also Landauer 1991)

This interpretation is also adopted by some philosophers of science; for instance, Peter Kosso states that "information is transferred between states through interaction" (1989: 37). The strong physical interpretation is the most common in the fields of physics and engineering: the need of a carrier signal sounds natural in the light of the generic idea that physical influences can only be transferred through interactions. Nevertheless, even in the context of this strong version, two different implicit ontological views can be distinguished.

According to a *substance view*, information belongs to the ontological category of substance, that is, it is an object of predication but not predicable of anything else, and a bearer of properties (see Robinson 2014). In this sense, the essential feature of information consists in its capability of "flowing" through the physical space, that is, of being generated at one point and transmitted to another point; it can also be accumulated, stored, and converted from one form to another. On the other hand, according to a *property view*, ontologically information is a property; in particular, it is a property of the carrier signal. Therefore, even if properties do not flow, the picture of the "flow" of information might make a certain sense: there is a propagation of the physical signal that links transmitter and receiver, and information is a property of that signal.

By contrast to a strong perspective, the *weak* physical interpretation does not require a physical interaction between two points of space to admit that there is transmission of information between them. By remaining agnostic about the ontological category of information, the only claim of the weak stance is that it is an item that belongs to the world of physics. In other words, the concept of information is not a merely formal concept: the term 'information' refers to something that is part of the furniture of reality as described by physical science. Therefore, even if the question of its ontological category is not decided yet, the behavior of information is governed by physical laws.

Although widespread in the physics and engineering communities, the strong physical interpretation faces a very serious challenge when entanglement-assisted communication comes into play. For instance, in teleportation an unknown quantum state is transferred between two points of space with the assistance of a shared pair prepared in an entangled state and of two classical bits sent from one point to the other. However, there is no physical signal that carries the teleported state. For this reason, many physicists try to find a physical link between the two physical points, for instance, by considering that information travels backward in time until the event at which the entangled pair was produced, and then travels forward to the future until the time in which the teleported state is recovered (Jozsa 1998, 2004; Penrose 1998), or by supposing that the information is carried by the classical bits, but in a hidden manner (Deutsch and Hayden 2000). If reluctant to accept this kind of artificial explanations, the advocate of the strong physical interpretation of information is in serious trouble. The weak physical perspective, on the contrary, is not challenged by entanglement-assisted communication, even though it is still in debt regarding the decision about the ontological category of information.

4.3 The Formal Interpretation

Although the physical interpretation has been the most usual in the traditional textbooks for engineers' training, this situation has changed in recent times: in general, present-day textbooks explain information theory in a formal way, with no mention of sources, receivers, or signals, and the basic concepts are introduced in terms of random variables, probability distributions over their possible values, and correlations between them (see, e.g., Cover and Thomas 1991). This trend can be viewed as implying a formal interpretation, according to which the term 'information' does not refer to something that exists in the world, but neither does it have to do with knowledge. Strictly speaking, the word belongs not to empirical science but to formal science: it has no extralinguistic reference in itself; its "meaning" has only a syntactic dimension. Therefore, information is a purely formal concept and the theory of information is a chapter of the theory of probability (see, e.g., Khinchin 1957; Reza 1961).

In a certain sense, the deflationary approach to information, defended by Timpson (2004, 2008, 2013) and Duwell (2008), can be viewed as a variant of the formal interpretation:

[O]nce it is recognized that 'information' is an abstract noun, then it is clear that there is no further question to be answered regarding how information is transmitted in teleportation that goes beyond providing a description of the physical processes involved in achieving the aim of the protocol. That is all that "How is the information transmitted?" can intelligibly mean; for there is not a question of information being a substance or entity that is transported, nor of 'the information' being a referring term.

(Timpson 2008: 599)

As a consequence, from this perspective, information theories (classical and quantum, and also computation theory) are "theories about what we can do using physical systems" (Timpson 2013: 69). Moreover, as explained previously, the success of communication is defined by means of the type-token distinction, which, in turn, is characterized in terms of a one-to-one mapping between states of the source and states of the destination (Duwell 2008) or of sameness of structure (Timpson 2013). Therefore, information transmission seems to be something that depends only on the formal features of the communicational arrangement, independently of how these features are implemented by physical means.

Of course, the formal approach is free from the difficulties that challenge the physical interpretation: if the word 'information' has no reference in the world, the problems of explaining how information is transferred in teleportation or of deciding the ontological category of information suddenly disappear. However, conceiving information as a merely formal concept leads to the same problem that stalks the epistemic interpretation: the magnitudes involved in the mathematical theory of information lose even their nomic ingredient. For instance, the mutual information between two random variables can be defined even if there is no lawful relationship between them and their conditional probabilities express only de facto correlations. Therefore, the formal interpretation does not incorporate the element of production implicit in communication, which permits that some action taken at the source of information has an effect on the destination of the information. In other words, although gaining in generality, the formal view of information loses conceptual soundness in the communicational context.

5 Communication and Quantum Information Theory

Although there were many works on the matter before Benjamin Schumacher's article "Quantum Coding" (1995) (see, e.g., Ingarden 1976), this work is usually considered the first precise formalization of quantum information theory. The main aim of the article is to prove a theorem for quantum coding analogous to the noiseless coding theorem of Shannon's theory. With this purpose, Schumacher conceives the message source A as a system of n states-letters a_i, each with its own probability $p(a_i)$; then, A has a Shannon entropy $H(A)$ computed as in eq. (1). In turn, the transmitter T maps the set of the states-letters a_i of the source A onto a set of n states $|a_i\rangle$ of a quantum system M. The states $|a_i\rangle$ belong to a Hilbert space $\mathcal{H}_M$ of dimension $\dim(\mathcal{H}_M) = d$ and may be non-orthogonal. The mixture of states of the signal source M can be represented by a density operator:

$$\rho = \sum_{i=1}^{n} p(a_i)|a_i\rangle\langle a_i| \qquad (9)$$

whose von Neumann entropy is:

$$S(\rho) = -\mathrm{Tr}(\rho \log \rho) \qquad (10)$$

In the case that the $|a_i\rangle$ are mutually orthogonal, the von Neumann entropy is equal to the Shannon entropy: $S(\rho) = H(A)$. In the general case, $S(\rho) \leq H(A)$.

Given this mapping, the messages $(a_{i1}, a_{i2}, \ldots, a_{iN})$ of N letters produced by the message source A are encoded by means of sequences of N quantum states $(|a_{i1}\rangle, |a_{i2}\rangle, \ldots, |a_{iN}\rangle)$, with $i \in \{1, 2, \ldots, n\}$. This sequence can be represented by the state $|\alpha\rangle = |a_{i1}, a_{i2}, \ldots, a_{iN}\rangle$ of a system M^N, belonging to a Hilbert space

$\mathcal{H}_{M^N} = \mathcal{H}_M \otimes \mathcal{H}_M \otimes \ldots \otimes \mathcal{H}_M$ (N times), of dimension d^N. This state is transmitted through a channel C composed of L two-state systems Q called *qubits*, each represented in a Hilbert space $\mathcal{H}_Q$ of dimension 2. Therefore, the Hilbert space of the channel will be $\mathcal{H}_C = \mathcal{H}_Q \otimes \mathcal{H}_Q \otimes \ldots \otimes \mathcal{H}_Q$ (L times), of dimension 2^L. Analogously to the Shannon case, L indicates the compactness of the code: the lower the value of L, the greater the efficiency of the coding, that is, fewer qubits are needed to encode the messages. The *Quantum Noiseless-Channel Coding Theorem* proves that, for sufficiently long messages, the optimal number $L_{\min}$ of qubits necessary to transmit the messages generated by the source with vanishing error is given by $NS(\rho)$.

Schumacher designs the proof of the theorem by close analogy with the corresponding Shannon's theorem. The idea is that all the possible states $|\alpha\rangle$ (representing the messages of N letters produced by the message source A), belonging to $\mathcal{H}_{M^N}$ of dimension $d^N = 2^{N \log d}$, fall into two classes: one of typical states belonging to a subspace of $\mathcal{H}_{M^N}$ of dimension $2^{NS(\rho)}$, and the other of atypical messages. When $N \to \infty$, the probability of an atypical state becomes negligible; so the source can be conceived as producing only messages represented by states belonging to a subspace of $2^{NS(\rho)}$ dimensions. Therefore, the channel can be designed to be represented in a Hilbert space $\mathcal{H}_C$ such that $\dim(\mathcal{H}_C) = 2^{NS(\rho)}$, and this means that the minimum number $L_{\min}$ of qubits necessary to transmit the messages of the source is $L_{\min} = NS(\rho)$.

Schumacher's formalism had a great impact on the physicist community: it is very elegant, and its analogy with Shannon's classical work is clear. Nevertheless, these facts do not yet supply an answer about the concept of quantum information. The positions about the matter range from those who seem to deny the existence of quantum information (Duwell 2003), those who consider that it refers to information when it is encoded in quantum systems (Caves and Fuchs 1996), and those who conceive it as a new kind of information qualitatively different from classical information (Jozsa 1998; Brukner and Zeilinger 2009). In this section, we will analyze the most common arguments in favor of this last position (for a detailed discussion, see Lombardi, Holik, and Vanni 2016a).

5.1 Quantum Sources of Information?

A usual claim is that quantum information is what is produced by a quantum information source, that is, a device that generates different quantum states with their corresponding probabilities (see, e.g., Timpson 2004, 2008, 2013; Duwell 2008). Those who adopt this characterization of quantum information in general stress the elegant parallelism between Shannon's and Schumacher's proposals.

A first point to notice is that this is not what Schumacher says. On the contrary, by following closely the terminology Shannon introduced through the entire paper, he begins by defining the *message source A* that produces each a_i with probability $p(a_i)$, and only in the stage of coding does he introduce the *quantum signal source*, which "is a device that codes each message a_M from the source A into a 'signal state' $|a_M\rangle$ of a quantum system M" (Schumacher 1995: 2738). This means that the quantum states involved in the process Schumacher described do not come from A, but from a quantum system M that is part of the device that encodes the *messages* produced by A and turns them into *signals* to be transmitted through the channel: M is part of the device called "transmitter." Schumacher calls the process developed between transmitter and receiver *'transposition'*:

We can therefore imagine a communication scheme based upon transposition. At the coding end, the signal of a source system M is transposed via the unitary evolution U into the coding system X. The system X is conveyed from the transmitter to the receiver. At the decoding end, the unitary evolution U' is employed to recover the signal state from X into M', an identical copy of system M.

(1995: 2741)

In this quote it is clear that the system *X is conveyed from the transmitter to the receiver*, not from the message source A to the message destination B. Moreover, the system M is placed at the *coding end* and the system M' is placed at the *decoding end*; so, M is not the message source A. In other words, the focus of the paper is on the stage of coding in the transmitter, transmitting through the channel, and decoding at the receiver, in agreement with what is suggested by the title itself of Schumacher's article: "Quantum Coding" and not "Quantum Information."

Nevertheless, it could be argued that, in spite of what Schumacher says, nothing prevents us from considering M a quantum source and defining quantum information as what is generated by a quantum source. But this position has additional difficulties.

First, this view implies confusing the effectiveness of communication, measured by the *mutual information* $H(A;B)$, with the effectiveness of transposition, measured by the *fidelity F* of the process, defined as (Schumacher 1995: 2742):

$$F = \sum_{i=1}^{n} p(a_i) \, \mathrm{Tr}|a_i\rangle\langle a_i|\omega_i \qquad (11)$$

where the $|a_i\rangle\langle a_i|$ correspond to the signal states produced at M, and the ω_i represent the signal states obtained at M' as the result of the transposition, which do not need to be pure (here we consider pure signal states, but the definition can be generalized to mixed signal states). Since fidelity measures the effectiveness of the stage of transmission through the channel, it is a property of the channel: the fidelity of a transmission is less than unity when the channel is limited in the sense that $\dim(\mathcal{H}_C) < \dim(\mathcal{H}_{M^N})$ (although indefinitely close to unity when $\dim(\mathcal{H}_C) = 2^{NS(\rho)}$, as proved by the quantum coding theorem). By contrast, communication is maximally effective when $H(A;B)$ is maximum, where $H(A;B)$ is a function not only of the fidelity of the transposition, but also of the reliability of the operations of coding and decoding, which correlate the states a_i of the message source A with the quantum states $|a_i\rangle$ of M, and the quantum states ω_i of M' and the states b_i of the destination B, respectively.

On the other hand, the very idea of a quantum source of information leads to conceptual perplexities. If the quantum states $|a_i\rangle$ to be transmitted through the channel were not the symbols by means of which the letters a_i of the message are encoded in the transmitter, but they were the elements of the message itself as produced by the quantum source of information, where would the coding process be located? In fact, what is produced by the information source would be the same as what is transmitted, and the term 'coding' would turn out to be vacuous: the title of Schumacher's article – as well as the quantum coding theorem – would lose its meaning.

If quantum information were fully identified with the quantum states produced by a quantum source, the transmission of information would be reduced to the transposition of quantum states. Indeed, if the fact that transposition is only a part of the communication process were forgotten and the roles played by the message source and the message destination were conflated, nothing would change in the discourse about quantum information if the term 'quantum information' were replaced by the term 'quantum state.' The argument can be posed in other terms: since quantum information is what is communicated and a quantum state is what is transposed, the identification between communication and transposition amounts to the identification between quantum information and quantum state. As Duwell clearly states, when the properties that supposedly motivate a new concept of information are revised,

[i]t is obvious that there is already a concept that covers all of these properties: the quantum state. The term 'quantum information' is then just a synonym for an old concept.

(2003: 498)

In other words, 'quantum information' turns out to mean quantum state, and the whole meaningful reference to communication gets lost.

5.2 Quantum Information and Quantum Coding

Another strategy to conceive quantum information as a different and peculiar type of information is to link the very meaning of the concept of information with the coding theorems: if the theorems are different in the classical case and the quantum case, the corresponding concepts of information are also different.

In Subsection 3.3, several caveats against defining information in terms of the coding theorems were presented. Those arguments also apply in this quantum domain. But in this case, that strategy has to face another conceptual challenge. In fact, if there were two qualitatively different kinds of information whose nature, classical and quantum, were defined by the corresponding coding theorems, then the source A would generate different kinds of information with no change in its own nature: the kind of information generated would depend not on itself, but on how the messages will be encoded later. Moreover, if the kind of coding to be used at the coding stage were not decided yet when the message is produced at the source A, the very nature – classical or quantum – of the information generated by A and carried by the message would be indeterminate, and would remain as such up to the moment of the decision.

Again, the difficulties disappear when the two concepts involved in communication are conceptually distinguished: on one hand, the information generated at the message source, which depends on the probability distribution over the source's states and is independent of coding – even independent of whether the messages are encoded – and, on the other, the resources necessary to encode the occurrence of those states, which also depend on the particular coding selected, classical or quantum.

5.3 *Quantum Information without Physical Carrier*

As explained in Subsection 4.2, the strong physical interpretation of communication, which requires a physical signal that carries the information between two points of space, faces a serious challenge when entanglement-assisted communication is considered, which establishes the quantum communicational link via the entangled state of a pair of particles, with no physical signal traversing space. But, why not consider precisely this feature that makes quantum information qualitative different from classical information? Whereas classical information always requires a physical carrier that travels through space in a finite amount of time, quantum information would not need a physical carrier to be transferred. Nevertheless, this view, although it would justify talking about quantum information in teleportation, is not free of difficulties.

First, although teleportation is a way of taking advantage of entanglement to implement transposition, this does not mean that any transposition process needs to be implemented by entanglement. Transposition only requires the signal to be conveyed from the transmitted to the receiver: "The system X is conveyed from the transmitter to the receiver. [It] supports the transposition of the state of M

into M' " (Schumacher 1995: 2741). It is clear that the process of transposition can be carried out by means of entanglement, in particular, of teleportation; but teleportation is only an example of the process: "'quantum teleportation' [...] is a rather exotic example of a transposition process" (Schumacher 1995: 2741). But transposition can also be met by sending the quantum physical system X from M to M' through space and time, and the whole formalism of quantum information theory still applies. Therefore, it is not communication without physical signal that characterizes the kind of information studied by quantum information theory.

In turn, the idea that the hallmark of quantum information is that it does not need a physical carrier to be transferred faces conceptual puzzles similar to those already pointed out in the previous subsection. Again, the source of messages A would generate different kinds of information with no change in its own nature, but now depending on whether the process of transposition will be later implemented by means of a carrier. Furthermore, the nature of the information generated at A would remain indeterminate until we decide to implement transposition by means of entanglement or by sending quantum systems through space from the transmitter to the receiver.

6 Concluding Remarks

In this chapter, we considered different issues regarding the notion of information, which, we believe, contribute to the elucidation of the concept in its several meanings.

We began by distinguishing the different ways in which the term 'information' can be used in very dissimilar fields. This task allowed us to focus on the mathematical concept of information in the communicational context, where Shannon's theory constitutes the classical formalism. In this formal framework, we discussed certain definitional points that can still be a matter of debate, and we have advocated a traditional view, according to which the – average – amount of information produced by the source is defined by the features of the source itself – the probabilities of its states – and is independent of coding, even of whether the messages are encoded.

We have then shown that, even with no formal disagreements, different interpretations of the concept of information coexist. In particular, we have considered the epistemic interpretation that links information with knowledge, the physical interpretation in its usually not sufficiently distinguished versions, and the formal interpretation that turns the concept of information into a mathematical concept. On this basis, we have shown that none of them is completely free of difficulties.

In the field of quantum information, we have argued that there are no sufficiently convincing reasons to accept that there is a kind of information, quantum information, qualitatively different from classical information. On this basis, interpreting information as theoretically neutral is perhaps more adequate in the communicational context. From this perspective, information is not classical or quantum, but it is neutral with respect to the physical theory that describes the systems used for its implementation. Many conceptual challenges simply vanish when it is assumed that the difference between the classical and the quantum cases is confined to the coding stage and does not affect the very nature of information.

Moreover, this theoretically neutral view of information has further conceptual advantages. On one hand, if the very concept of quantum information is unavoidably linked to quantum mechanics, the program of deducing quantum mechanics from the constraints imposed by quantum information runs the risk of becoming circular. The risk of circularity is avoided by accepting that there is a single kind of information, which is not tied to a particular physical theory.

On the other hand, although unification in science has been widely considered a desirable goal, at present reductionism is viewed with skepticism in both the physical and the philosophical communities. The theoretical neutrality of information makes it possible to preserve the ideal of unification without commitment to reductionism. If different physical theories can be reconstructed on the same neutral informational basis, they could be meaningfully integrated into a single theoretical network and compared to each other, with no need to search for reductive links among them.

Finally, the theoretically neutral way of conceiving information paves the way to consider the prolific present-day research about classical models for quantum information. To what extent is quantum mechanics necessary to implement the protocols of quantum information theory? Is the whole theoretical power of quantum mechanics necessary in every case, or can certain quantum information processes be implemented by classical models? Are the peculiarities of quantum mechanics required for the possibility of implementing the information protocols, or only to obtain efficient implementations? These questions open a wide field of conceptual and philosophical research based on recent theoretical and experimental results. However, such work would not find a comfortable place in the framework of a position that presupposes the qualitative difference between classical and quantum information from the very beginning. On the contrary, the view of information as theoretically neutral is particularly adequate for undertaking that conceptual research: it has the sufficient flexibility to accommodate different ways of implementing informational processes, to the extent that they are not tied a priori to a particular physical theory.

These conclusions do not intend to underestimate the relevance of so-called quantum information theory, a field that has grown dramatically in recent decades, supplying many new and significant results with promising applications. But from our perspective, the theory does not deal with a kind of information qualitatively different from classical information, but is concerned with the application of quantum resources to information theory.

Acknowledgments

We are extremely grateful to all the participants in the workshop "What Is Quantum Information?" (Buenos Aires, May 2015): Jeffrey Bub, Adán Cabello, Dennis Dieks, Armond Duwell, Christopher Fuchs, Robert Spekkens, and Christopher Timpson, for the stimulating and lively discussions about the concept of information. This chapter was partially supported by a Large Grant of the Foundational Questions Institute (FQXi), and by grants of the National Council of Scientific and Technological Research (CONICET) and the National Agency for Scientific and Technological Promotion (ANPCyT-FONCYT) of Argentina.

References

Adriaans, P. (2013). "Information." In E. N. Zalta (ed.), *The Stanford Encyclopedia of Philosophy* (Fall 2013 Edition), http://plato.stanford.edu/archives/fall2013/entries/information/.

Adriaans, P. and Van Benthem, J. (2008). *Philosophy of Information (Handbook of the Philosophy of Science 8)*. Amsterdam: North-Holland.

Bar-Hillel, Y. (1964). *Language and Information: Selected Essays on Their Theory and Application*. Reading, MA: Addison-Wesley.

Bar-Hillel, Y. and Carnap, R. (1953). "Semantic Information." *The British Journal for the Philosophy of Science*, **4**: 147–157.

Barwise, J. and Seligman, J. (1997). *Information Flow: The Logic of Distributed Systems*. Cambridge: Cambridge University Press.

Bell, D. (1957). *Information Theory and Its Engineering Applications*. London: Pitman & Sons.

Brukner, Č. and Zeilinger, A. (2009). "Information Invariance and Quantum Probabilities." *Foundations of Physics*, **39**: 677–689.

Bub, J. (2007). "Quantum Information and Computation." Pp. 555–660 in J. Butterfield and J. Earman (eds.), *Philosophy of Physics. Part B*. Amsterdam: Elsevier.

Caves, C. M. and Fuchs, C. A. (1996). "Quantum Information: How Much Information in a State Vector?" In A. Mann and M. Revzen (eds.), *The Dilemma of Einstein, Podolsky and Rosen – 60 Years Later. Annals of the Israel Physical Society*, **12**: 226–257 (see also quant-ph/9601025).

Caves, C. M., Fuchs, C. A., and Schack, R. (2002). "Unknown Quantum States: The Quantum de Finetti Representation." *Journal of Mathematical Physics*, **43**: 4537–4559.

Chaitin, G. (1966). "On the Length of Programs for Computing Binary Sequences." *Journal of the Association for Computing Machinery*, **13**: 547–569.

Cover, T. and Thomas, J. (1991). *Elements of Information Theory*. New York: John Wiley & Sons.

Deutsch, D. and Hayden, P. (2000). "Information Flow in Entangled Quantum Systems." *Proceedings of the Royal Society of London A*, **456**: 1759–1774.

Dretske, F. (1981). *Knowledge & the Flow of Information*. Cambridge, MA: MIT Press.

Dunn, J. M. (2001). "The Concept of Information and the Development of Modern Logic." Pp. 423–427 in W. Stelzner (ed.), *Non-classical Approaches in the Transition from Traditional to Modern Logic*. Berlin: de Gruyter.

Duwell, A. (2003). "Quantum Information Does Not Exist." *Studies in History and Philosophy of Modern Physics*, **34**: 479–499.

Duwell, A. (2008). "Quantum Information Does Exist." *Studies in History and Philosophy of Modern Physics*, **39**: 195–216.

Fetzer, J. H. (2004). "Information: Does It Have to Be True?" *Minds and Machines*, **14**: 223–229.

Floridi, L. (2004). "Outline of a Theory of Strongly Semantic Information." *Minds and Machines*, **14**: 197–222.

Floridi, L. (2005). "Is Information Meaningful Data?" *Philosophy and Phenomenological Research*, **70**: 351–370.

Floridi, L. (2010). *Information – A Very Short Introduction*. Oxford: Oxford University Press.

Floridi, L. (2011). *The Philosophy of Information*. Oxford: Oxford University Press.

Floridi, L. (2015). "Semantic Conceptions of Information." In E. N. Zalta (ed.), *The Stanford Encyclopedia of Philosophy* (Spring 2015 Edition), http://plato.stanford.edu/archives/spr2015/entries/information-semantic/.

Graham, G. (1999). *The Internet: A Philosophical Inquiry*. London: Routledge.

Hartley, R. (1928). "Transmission of Information." *Bell System Technical Journal*, **7**: 535–563.

Hoel, E., Albantakis, L., and Tononi, G. (2013). "Quantifying Causal Emergence Shows that Macro Can Beat Micro." *Proceedings of the National Academy of Sciences*, **110**: 19790–19795.

Ingarden, R. (1976). "Quantum Information Theory." *Reports on Mathematical Physics*, **10**: 43–72.

Jozsa, R. (1998). "Entanglement and Quantum Computation." Pp. 369–379 in S. Huggett, L. Mason, K. P. Tod, S. T. Tsou, and N. M. J. Woodhouse (eds.), *The Geometric Universe*. Oxford: Oxford University Press.

Jozsa, R. (2004). "Illustrating the Concept of Quantum Information." *IBM Journal of Research and Development*, **4**: 79–85.

Khinchin, A. (1957). *Mathematical Foundations of Information Theory*. New York: Dover.

Kolmogorov, A. (1965). "Three Approaches to the Quantitative Definition of Information." *Problems of Information Transmission*, **1**: 4–7.

Kolmogorov, A. (1968). "Logical Basis for Information Theory and Probability Theory." *Transactions on Information Theory*, **14**: 662–664.

Kosso, P. (1989). *Observability and Observation in Physical Science*. Dordrecht: Kluwer.

Landauer, R. (1991). "Information Is Physical." *Physics Today*, **44**: 23–29.

Landauer, R. (1996). "The Physical Nature of Information." *Physics Letters A*, **217**: 188–193.

Lombardi, O. (2004). "What Is Information?" *Foundations of Science*, **9**: 105–134.

Lombardi, O. (2005). "Dretske, Shannon's Theory and the Interpretation of Information." *Synthese*, **144**: 23–39.

Lombardi, O., Fortin, S., and López, C. (2016). "Deflating the Deflationary View of Information." *European Journal for Philosophy of Science*, **6**: 209–230.

Lombardi, O., Fortin, F., and Vanni, L. (2015). "A Pluralist View about Information." *Philosophy of Science*, **82**: 1248–1259.

Lombardi, O., Holik, F., and Vanni, L. (2016a). "What Is Quantum Information?" *Studies in History and Philosophy of Modern Physics*, **56**: 17–26.

Lombardi, O., Holik, F., and Vanni, L. (2016b). "What Is Shannon Information?" *Synthese*, **193**: 1983–2012.

MacKay, D. (1969). *Information, Mechanism and Meaning*. Cambridge MA: MIT Press.

Nauta, D. (1972). *The Meaning of Information*. The Hague: Mouton.

Penrose, R. (1998). "Quantum Computation, Entanglement and State Reduction." *Philosophical Transactions of the Royal Society of London A*, **356**: 1927–1939.

Reza, F. (1961). *Introduction to Information Theory*. New York: McGraw-Hill.

Robinson, H. (2014). "Substance." In E. N. Zalta (ed.), *The Stanford Encyclopedia of Philosophy* (Spring 2014 Edition), http://plato.stanford.edu/archives/spr2014/entries/substance/.

Scarantino, A. and Piccinini, G. (2010). "Information without Truth." *Metaphilosophy*, **41**: 313–330.

Schumacher, B. (1995). "Quantum Coding." *Physical Review A*, **51**: 2738–2747.

Shannon, C. (1948). "The Mathematical Theory of Communication." *Bell System Technical Journal*, **27**: 379–423, 623–656.

Shannon, C. and Weaver, W. (1949). *The Mathematical Theory of Communication*. Urbana and Chicago: University of Illinois Press.

Solomonoff, R. (1964). "A Formal Theory of Inductive Inference." *Information and Control*, **7**: 1–22, 224–254.

Stonier, T. (1990). *Information and the Internal Structure of the Universe: An Exploration into Information Physics*. New York-London: Springer.

Stonier, T. (1996). "Information as a Basic Property of the Universe." *Biosystems*, **38**: 135–140.

Timpson, C. (2004). *Quantum Information Theory and the Foundations of Quantum Mechanics*. PhD diss., University of Oxford (arXiv: quant-ph/0412063).

Timpson, C. (2006). "The Grammar of Teleportation." *The British Journal for the Philosophy of Science*, **57**: 587–621.

Timpson, C. (2008). "Philosophical Aspects of Quantum Information Theory." Pp. 197–261 in D. Rickles (ed.), *The Ashgate Companion to the New Philosophy of Physics*. Aldershot: Ashgate Publishing (page numbers are taken from the online version: arXiv: quant-ph/0611187).

Timpson, C. (2013). *Quantum Information Theory and the Foundations of Quantum Mechanics*. Oxford: Oxford University Press.

Zeilinger, A. (1999). "A Foundational Principle for Quantum Mechanics." *Foundations of Physics*, **29**: 631–643.

2

Representation, Interpretation, and Theories of Information

ARMOND DUWELL

1 Introduction

On the first page of Shannon's (1948) paper there are a few lines that are often quoted as providing a starting point for interpreting Shannon's theory of information:

> The fundamental problem of communication is that of reproducing at one point either exactly or approximately a message selected at another point. Frequently the messages have meaning. ... These semantic aspects of communication are irrelevant to the engineering problem.

It is entirely appropriate to take Shannon's words of wisdom seriously, that the theory has nothing to do with the meaning of messages. That said, it is a mistake to dismiss all intentional concepts while interpreting the theory. In fact what I will demonstrate in this chapter is that focusing on the representational and interpretive aspects of theories is vitally important for understanding the definition of information provided in Timpson (2013), and how Shannon information theory and quantum information theory are related. These aspects have heretofore not been sufficiently appreciated with respect to the framework for understanding information theories developed by Timpson (2004, 2013) and Duwell (2008). I will be correcting that problem in this chapter, and arguing that information theory in general, and Shannon and quantum information theories in particular, involve far more conventional aspects than has been realized, and explain why they do so.

In order to appreciate the pragmatic aspects of communication, it is essential to draw one's attention to some basic features of representation and interpretation, which is the focus of Section 2. In Section 3, I will present the relevant aspects of Shannon's theory to bring out the features that are so important to understanding what information is. This section also raises a problem for the definition of information and proposes a resolution to the problem. The problem will illustrate how important users of Shannon's theory are in determining what constitutes successful Shannon information transfer. In Section 4, I will outline quantum information theory and raise another problem for the definition of information. Again, the resolution to the problem will depend on recognizing just how important users of quantum information theory are in determining what constitutes successful quantum information transfer. The resolution to the problem will also illustrate important differences between Shannon information theory and quantum information theory.

2 Representation and Interpretation

In this section I will discuss some basic features of representation and interpretation of theories that are especially salient for clarifying what information is, and how Shannon and quantum information theories are related. I will first discuss representation, and then interpretation.

2.1 Representation

Perhaps the most oft-cited work on representation is Goodman (1968). In part it is because he made the stunning claim that representation had nothing, at least essentially, to do with resemblance. Importantly, it is also because he was right. More important for representation was the symbol system that a representation was associated with, which determined what it represented. Regarding symbol systems, Goodman had few restrictions on them, save for the fact that they have associated syntactics and semantics. Natural languages certainly qualify, but symbol systems do not have to be in common usage to qualify as a symbol system. As such, it would seem that just about anything goes, and later developers of the theory of scientific representation, such as Hughes (1997), Suarez (2004), and Van Fraassen (2008), have said as much.

Van Fraassen writes:

There is no representation except in the sense that some things are used, made, or taken, to represent some thing as thus or so.

(2008: 23)

For the purposes of this chapter, I assume that all it takes to be a representation is to be used, made, or taken to be one. Given the extreme flexibility in what counts as a representation, no claims about *accuracy* of representation follow from *being* a representation. Of course, in science, we are interested in accurate representations, those that allow us to infer truths about representational targets, and the more the better. So, while representation is extremely permissive, successful scientific representation is not.

Van Fraassen characterizes theories as providing a space of possibilities. To enable a theory's representational resources, the user of a theory needs to locate the systems that he or she wants to represent in the space of possibilities that a theory provides. Put more prosaically, if you want to get everything out of a theory that you can, you need to feed in some initial or boundary conditions. Typically one goes about locating systems in the space of possibilities a theory countenances by making measurements associated with those possibilities.

Van Fraassen is at pains to emphasize that measurements alone are not enough. In order to actually *use* a theory to represent the world, one has to *judge* where one is in the space of possibilities that a theory countenances. Of course, one can and should use any pertinent information in arriving at such judgments. Measurements will be a part of the story, but they will not be all of the story. One needs to make judgments about whether the measurements made were an accurate (partial) representation of the world. To illustrate, Van Fraassen invites us to use a map even when there is a "you are here" spot marked on it. Such a spot will not enable the map's user to navigate if she doesn't agree that she is indeed located at the location represented by the spot on the map. Similarly, even if a measurement suggests that the mass of a particle is such and such, it doesn't automatically make it true. One's measurement device might have accidentally been damaged in an accident, and one might wonder if its characterization of a system of interest is useful.

One with reasonable realist inclinations might object to this indexical aspect of scientific representation. Aren't there some facts of the matter about what the properties of the system of interest are? In keeping with his constructive empiricism, Van Fraassen is noncommittal, but he has non-skeptical arguments for why one should be noncommittal on such matters. Characterizing a projectile as having a certain mass distribution seems like no great leap of faith, but the reason that it is no great leap of faith is because it is part of a historical process of development of both theory and measurement that has led to a certain kind of stability. The stability in question lies in the predictions of a theory being true, given a certain acceptable set of procedures for making the relevant measurements and for locating systems of interest in the space of possibilities the theory countenances. In his book, Van Fraassen has an extended discussion of the development of measurements for temperature and associated difficulties. Van Fraassen describes the ideal gas law as playing a central role in discovering appropriate measurements of temperature that were otherwise elusive. That said, it may have been the case that different measurements of temperature were taken as revealing important features of phenomena, and perhaps something besides the ideal gas law would have been successful as a result. Stability between theory and measurement is an epistemically secure fact for Van Fraassen; he is agnostic about what features of the world secure that stability. The take-home point is that there may be many different ways to secure stability when there are multiple moving parts (theory and measurement procedures). There is no way to remove the judgment of a user of a theory to enable its use.

It is not important to the overall argument of this chapter whether one adopts the views of representation discussed earlier. The discussion of the view is included not primarily to advocate for it, but rather to cognitively prime the reader to appreciate certain features of information theories that have hitherto gone unnoticed.

2.2 Interpretation

Ruetsche (2011) provides both a useful framework for thinking about interpreting theories and some useful advice about how to go about it. I will follow her in this regard.

Many physical theories can be viewed as providing a certain kind of structure that composes the representational resources of the theory. The structure a theory has can often be described as having three different substructures: a state space, a set of observables on the state space, and dynamics that indicate how states change over time. Describing the structure as such is an exercise in interpretation, but one that is philosophically uncontroversial. As Ruetsche emphasizes, physical theories come already partially interpreted. They are interpreted in such a way as to render a theory empirically applicable, and often empirically successful. The more philosophically controversial interpretational exercise is to specify under what conditions models of such substructures are instantiated. Hence, Ruetsche characterizes interpretation as an exercise in nomic articulation: to characterize the worlds possible according to the theory.

A reasonable guide to this interpretational exercise is to interpret theories so that they can be described as having the virtues realists are wont to attribute to approximately true or true theories. They should be consistent internally and with other theories. They should be fruitful in their application to known problems as well as suggest and provide resources for solving undiscovered problems. They ought to enable a theory's explanatory resources in a way that accounts for the successful applications of a theory as well.

One ramification of this advice on how to interpret theories is that interpretational monism is preferred to pluralism. A realist would be hard pressed to judge a theory as approximately true if one required different interpretations of it to enable its explanatory abilities. Of course, as Ruetsche details in her book, such cases exist in physics as currently practiced. Fortunately, there are no reasons to think that this is the situation with certain theories of information, e.g., Shannon's theory and quantum information theory, as I will demonstrate later.

The link between interpretation and representation that I would like to emphasize is that characterizing the worlds possible according to a theory will place constraints on what types of systems can be reasonably located in the space of possibilities that a theory countenances. Following Ruetsche's advice, this means that one should only locate systems in the space of possibilities that a theory countenances only if doing so maintains the success of the theory. This crucial constraint on locating systems in a theory will be the key to understanding what information is, and how Shannon and quantum information are related.

3 Shannon Information

I am now in a position to put this discussion on interpretation and representation to work to help to interpret Shannon's theory. In Section 3.1, I will discuss the indexical elements of Shannon's theory. In Section 3.2, I will discuss the definition of information. I will argue that the indexical character of Shannon's theory helps us understand how to interpret it. In Section 3.3, I will consider a problem for the definition and provide a solution to the problem. The solution will reveal that conventional aspects of information are inescapable.

3.1 Representation and Interpretation of Shannon's Theory

In this subsection I will briefly discuss the structure of Shannon's theory in relation to the discussion in the previous section. I will draw attention to the lack of constraints on locating systems in Shannon's theory. It will be this feature that allows for a great deal of convention, and hence input from users of the theory.

The great success of Shannon's (1948) theory is that it determines the best rate of communication given the statistical properties of a communication system. The best rate of communication is in part determined by the minimal number of code words required to distinguish any message from the source. Put differently, Shannon's theory characterizes the degree to which messages from the information source can be compressed given its statistical properties alone, and then determines the rate at which these compressed messages can be transferred over a channel.

Shannon's theory can be characterized as being composed of the theoretical substructures mentioned earlier. The state space of the theory is composed of all possible messages from all possible finite alphabets and possible associated probability distributions. This is a massively redundant structure. The compression of messages that is achievable is determined solely from the probability distribution associated with the communication system, not anything about the particular alphabet in use. So, any alphabet is as good as any other with respect to characterizing the information source with respect to Shannon's theory alone. Of course, some alphabets might be more convenient to use than others, but that is something determined by users of the theory.

The sense in which Shannon's theory can be characterized as having observables and dynamics is impoverished in comparison to many physical theories, but it is still possible. The observables associated with the theory can be represented by indicator functions for messages. The dynamics of the theory is even more impoverished. Shannon's theory requires the user of the theory to provide the statistical properties of all components of the communication system. It places no constraints on what these are, only on compressibility of messages and rates of communication given the statistical properties. So, the only sense of dynamics that can be recovered is that given a particular message from the source, one can, via conditional probabilities of the components of the communication system, provide the probability that the message will be reproduced at the destination. Note that absolutely no work is done by Shannon's theory here. The work is done by the user-provided statistical properties of the communication system.

In richer theories, the state space, observables, and dynamics are much more tightly interrelated than they are in Shannon's theory. Given a particular state, the observables tell us something about what properties a system has, or at least what we should expect from measurements of those observables. Given what we observe, we can often infer a state. Dynamics tells us how states change over time, and hence also how what we observe will vary over time. Given information about any two of the components, what the state is, what the dynamics is, or what the observables are, one can make inferences about the remaining component. So, these components of a theory mutually constrain each other.

The state space of Shannon's theory has very little structure from a mathematical point of view; it is just a set with a probability distribution on it. Different members of the set are completely unrelated to one another, save for their probabilities needing to add up to one. Correspondingly, the observables for different states will be completely unrelated to one another, again save for the probabilities to mirror the states they are observables of. As mentioned previously, the dynamics of a communication system is unconstrained by the theory. This lack of constraints in Shannon's theory has ramifications for what it can be used to represent.

Precisely because of the lack of structure in the state space of Shannon's theory, almost anything that can be described by a probability distribution can be considered a Shannon information source, subject to one important constraint. In order for Shannon's theory to make correct predictions about optimum communication rates, the systems or properties of the systems modeled as messages must be *distinguishable* from each other. As a result, the properties that qualify a system for being representable as one message can be completely unrelated to the properties that qualify a system as being representable as a different message. This indicates incredible flexibility as to what can count as an information source. It also has the consequence that the observables associated with certain letters or messages from a Shannon information source can be completely unrelated to one another.

There are no empirical regularities that are appealed to that determine whether a system instantiates one message as opposed to another; at least none imposed *by Shannon's theory*. It is completely conventional. Once a convention is stipulated by a user, there may be very precise conditions that qualify a system as instantiating one message rather than another, but *there are none without convention*. Similar claims about characterizing a system as a particular information source or other component of a communication system can be made with a caveat. In these cases frequencies are used to determine the appropriate probabilities to assign to letters of an alphabet. Empirical regularities are always important in assigning probabilities.

Consider an example of a simple communication system. A coin flip will be used as the information source. There are two choices involved in representing a coin flip as an information source. The first is to choose the alphabet, and the second is to choose which properties of the coin will correspond to the letters of the alphabet chosen. A very natural alphabet to use is {H, T}, corresponding to heads and tails in the obvious way. Once the alphabet and the properties associated with each letter are fixed, one can determine the appropriate probabilities to use to characterize the information source in Shannon's theory. That said, there is nothing about the physical system that is the information source that forces one to locate the system in that way in Shannon's theory. One might choose to use the alternative alphabet $\{H_N, H_S, T_N, T_S\}$ where the subscripts are meant to specify the orientation of the coin in terms of north or south. If a coin actually has a depiction of a head on it, the head can indicate the orientation of the coin. Once this alphabet and the properties associated with each letter are fixed, then the appropriate probabilities can be fixed and then the system can be properly described as an information source in Shannon's theory. Or one might choose a different alphabet {N, E, S, W} where one doesn't care which side of the coin is up, but one only cares about whether the coin is oriented north, east, south, or west. One can continue this line of reasoning ad nauseum. The only constraint on locating systems in Shannon's theory is that different states are associated with different systems that are distinguishable from one another. Else, nearly anything goes.

The upshot of this section is that the properties of a physical system, save for whether it is distinguishable or not, do not determine its description in Shannon's theory. It is the user of Shannon's theory who makes that determination and in doing so establishes a convention for his or her application of Shannon's theory. This is the indexicality of scientific representation in a rather pure form. These considerations must be held firmly fixed in mind in order to understand what information is.

3.2 Defining Information

In this subsection I will be discussing the definition of information provided by Timpson (2013). I will argue that this definition implies that there are always important conventional elements involved in communication.

Timpson writes:

> Information$_t$ is what is produced by an information$_t$ source that is required to be reproducible at the destination if the transmission is to be counted as a success.
> (2013: 22)

The subscript t is meant to emphasize that this definition concerns a particular kind of technical concept of information, the one associated with Shannon's theory as well as quantum information theory, in contrast to the everyday concept of information.

The definition raises interpretive questions: What is the relation between *what is produced at the information source* and *what needs to be reproducible for transmission to be counted as success*? Does what is produced by the information source determine success criteria for transmission, or do success criteria determine what is produced at the source? What is produced by an information source is an objective feature of the world completely independent from us, so that would seem to indicate that it determines the appropriate success criterion. Yet what is produced by the information source can come under many descriptions, and that is something that we have complete control over, so, perhaps it is better to say that the success criterion determines what is produced by the information source. Or is something more subtle going on?

It is natural to turn to Shannon's theory for help in adjudicating the appropriate interpretation. There is a natural criterion of successful transmission given the structure of Shannon's theory, especially its state space: the message produced has to be the message received. So, if a system is represented in the theory by a sequence from an alphabet, *abfsdn*, then the system received at the destination must also be appropriately represented by the same sequence, *abfsdn*. That said, as discussed earlier, Shannon's theory places no restrictions whatsoever on what systems can be represented by alphabetic sequences like *abfsdn*, save that they are distinguishable from other sequences. So, because the structure of Shannon's theory places almost no constraints on what states represent, this will not indicate anything about whether what is produced by the information source determines success criteria or the other way around.

Perhaps there are restrictions placed on how to locate systems in Shannon's theory that don't concern the particular structure of the theory, but concern the definition of information. If information is a *type* of thing, and a type of thing is something that needs to be reproduced for successful communication, the type is what determines which systems can appropriately be represented as the same alphabetic sequence in the theory. As Duwell (2008) has pointed out, there is no *unique* type associated with any physical system. We can characterize an information source as producing tokens of a particular type, but there is no unique token of a type that it produces. It produces tokens of many types. The example cited previously of a coin toss is useful to keep in mind here. A particular toss of a coin can be viewed as a token of the type *coin coming up heads*, or of the type coin *coming up heads and oriented in a northerly direction*, or of the type *oriented in a northerly direction*. Nothing singles out which of these types to focus on. So, the type, or more aptly, types of things produced by the information source *cannot* determine success criteria for communication. It is here where conventional considerations come into play.

I asked whether the type of thing produced by the information source determines what constitutes successful communication or whether the success criteria for successful communication determine what types of thing the information source produces. The dichotomy is misleading; the types produced by an information source and success criteria go hand in hand. A user can provide success criteria for communication, and, via the definition of information, indirectly specify the types of interest by indicating what counts as tokens of those types. Alternatively, a user can specify the types of interest for communication, and via the definition of information and identity conditions for those types provide success criteria. There is an equivalence of sorts between the two. What is crucial is the essential role a user of a communication system plays in communication systems. Users stipulate an appropriate conventional choice either of types to focus on or success criteria.

3.3 A Problem for the Definition of Information

In keeping with Ruetsche's advice about how to go about interpreting theories, one should seek an interpretation of Shannon's theory as well as the concept of information it employs in a way to preserve the success of the theory. In this subsection I will consider an example of a successful application of Shannon's theory that appears to conflict with the interpretation of the definition of information offered earlier. In order to solve this problem one needs to recognize that information is even more conventional than realized thus far.

Consider a commonplace example of a control system, a video game system. The user inputs commands into the controller, and the system responds by certain actions happening on a screen. One can view control systems as communication systems. The information source is the controller and the destination is the screen. Certain controller sequences correspond to certain actions. Shannon's theory will specify the minimal resources required to differentiate different control sequences and hence communicate them to the destination. In keeping with the definition of information previously introduced, communication is successful when the same type produced at the information source is token instantiated at the destination. This means that commands have to be viewed *as the same type* as actions. This might strike one as a category mistake. How can commands and actions be the same type of things?

The answer is simply this: pure stipulation or convention. The existence of purely conventional types is ubiquitous as evidenced by the writing on this page. All linguistic types are conventional. While languages have a history and conventions can be firmly fixed, such facts only give the user an illusion that tokens of a type *necessarily* share some physical properties in common. That many do is no evidence that they must. Morse code is an especially useful example. By an act of stipulation certain sequences of sounds, completely unrelated to the way English is actually spoken, became tokens of letters in English. What is confusing about the video game example is that the particular systems that form the information source and destination can be described as being tokens of very different kinds of types: commands and actions. Insofar as the very same system can be a token of many different types simultaneously, there is no contradiction here. Tokens of words can come in the form of sounds or inscriptions. Sounds and inscriptions are incompatible types, but that doesn't prevent tokens of these types from being tokens of words.

In order to be compatible with the successful application of Shannon's theory, the definition of information must be understood so that the type of thing produced by an information source can be given by convention. Tokens of a particular type need share almost no properties in common except being tokens of the same type. So, conventional judgments come into the theory in two different ways. A user of a Shannon communication system chooses which of the myriad of types associated with the output of an information source to focus on. Additionally, of those types, one needs to recognize that they can be completely conventionally defined. Of course, conventions, especially for isolated use, come cheap. This allows for the tremendous flexibility of application of Shannon's theory. As we will see, quantum information theory has a great deal of convention associated with it as well.

4 Quantum Information

In this section I will use insights developed earlier in this chapter to investigate quantum information theory. I want to emphasize that I am not interested in information-theoretic interpretations of quantum mechanics in this section.

Instead, I will focus on the theory that originated in Schumacher (1995) and its developments. I am interested in how to interpret the concept of quantum information, and how it relates to Shannon information. In Section 4.1, I will discuss the indexical features associated with quantum information theory. In Section 4.2, I will discuss the definition of quantum information. In Section 4.3, I will raise a problem for the definition of quantum information that will illustrate the role of convention in quantum information theory and clarify how to understand the differences between Shannon information and quantum information.

4.1 Representation and Interpretation of Quantum Information Theory

The aim of quantum information theory is to describe the best rate at which quantum messages can be communicated given the quantum and statistical properties of a communication system alone. Part of what determines the best rate of communication is the minimal dimension of Hilbert space required to reconstruct a quantum message. Put differently, quantum information theory characterizes how much messages from the information source can be compressed given their quantum and statistical properties.

Quantum information theory is an application of standard quantum mechanics. It can be characterized as being composed of a state space, a set of observables, and dynamics. The state space of the theory is given by the set of density operators acting on a Hilbert space, the observables are the set of positive operator value measures on that space, and the dynamics are given by the set of linear, trace-preserving, completely positive maps on operators. The state space and observables associated with the theory are of special importance for understanding the theory. This is a much richer structure than that associated with Shannon's theory, and it has important ramifications.

Hilbert spaces are vector spaces equipped with an inner product. The inner product relates different possible states, which is a marked difference from the state space of Shannon's theory. States can be judged to be closer or further away from one another because of this inner product.

Quantum messages are composed of sequences of systems in particular quantum states. A quantum information source outputs systems in quantum states according to a probability distribution. It might output sequences of systems in pure or mixed states. Successful transfer of quantum messages consists in reproducing a system at the destination that behaves, in certain respects and degrees, like the system produced by the quantum information source.

Quantum information theory has a much different relation to measurement than Shannon's theory. Measurements of observables are how one locates a particular system in the space of possibilities countenanced by a theory. One can make measurements to determine an appropriate quantum state to assign to a system according to a particular preparation. One can make measurements on a system after undergoing a particular preparation to determine the dynamics associated with a particular experimental setup. Measurements in this case will determine a final state, and one can work back to the appropriate dynamics. In other circumstances, one might have a firm grasp on the dynamics, but be uncertain about what the initial state of a system is. By making measurements on the final state, one can infer the initial state. One might also have a firm fix on an initial state and an ability to alter the dynamics and use that to determine what observable is associated with a particular measurement device. Quantum information theory and its parallel experimental procedures have achieved stability such that the predictions of the theory match what is expected on the basis of experimental procedure.

Measurement techniques and theory typically evolve together until stability is reached between results of measurements and predictions of a theory. This is certainly the case in quantum information theory. Unlike the case of Shannon's theory, there *are* empirical regularities that are appealed to that can justify that a system instantiates one message as opposed to another. What states were produced by an information source is in this regard *not* completely conventional. This isn't to say that no conventions are associated with the theory. Like any other theory, a coordinate system must be stipulated in order to generate any predictions, and so on.

Despite the fact that there is stability between theory and experimental procedure, the indexical character of the theory is present. In an indirect way, no one has brought this to our attention better than Christopher Fuchs and his collaborators (see Caves et al. 2007, or Timpson 2008 for synopses of the overall view). He has relentlessly reminded us how much judgments are a part of quantum mechanics. One's judgments about what state a quantum system is in are inexorably tied to other judgments. One might think that this isn't the case. If a measurement is performed on a quantum system, then there is a straightforward rule that indicates what the appropriate state assignment is. That being so, in order to actually locate the system in the space of possibilities quantum information theory countenances, one has to make a judgment about what observable was measured during the measurement procedure. Of course, one makes *those* judgments by appealing to an additional mix of empirical and theoretical considerations. There are three moving pieces of quantum information theory: states, observables, and dynamics. In order to secure the empirical success of the theory, judgments about one's location in theoretical space have to be appropriately coordinated between these three pieces. Any systematic differences between judgments of users of the theory regarding, say, state assignments can be systematically compensated for by systematically different judgments regarding dynamics or observables. This isn't to say that things are willy-nilly. If one's judgments depart too much from those justified by the hard-won knowledge of how to coordinate theory and experimental procedure, the empirical success of the theory will be compromised. The difference in constraints between Shannon information theory and quantum information theory marks an important difference between the two theories, one that will play an important role in understanding the differences between Shannon information and quantum information.

4.2 Defining Quantum Information

In Section 3.2, I examined whether Shannon's theory placed any natural structural constraints on the types of things that could count as information. The natural structural constraint was that messages from the information source can be assigned the same state as messages at the destination. That said, because there are so few constraints on locating systems as particular messages, this left the types of things that could count as Shannon information open to convention. In this subsection, I want to pursue the same strategy with respect to quantum information theory.

Quantum information theory provides a natural structural constraint on what constitutes successful information transfer. Duwell (2008) has advocated high entanglement fidelity as the appropriate internal success criterion for quantum information theory. Entanglement fidelity is a measure of how close an initial state is to that state after it has undergone some kind of process. High entanglement fidelity guarantees that the initial and final states will have high probability of passing any test of sameness. Importantly, that means not only will the probability distributions associated with *individual* observables of the quantum message be preserved in quantum information transfer, but also *joint* observables that reveal entanglement that the quantum message might have with other systems. This marks a striking difference between Shannon's theory and quantum information theory. This difference is precisely because quantum mechanics admits of non-local states, where the reduced state of the individual components cannot be used to recover the joint state. Certainly no distinction is made between individual and joint states in Shannon's theory. The natural structural success criteria of the two theories are fundamentally different, which can be obscured by imprecise language when we describe them as having the same message at the destination as was produced at the source.

As mentioned previously, the natural structural success criterion on Shannon's theory places very few constraints on the types of systems that can qualify as instances of Shannon information because there are so few constraints on state assignments. The same is not true of quantum information theory. Because of the coordination between experimental procedure and theory, one cannot go assigning states to systems willy-nilly because it will not preserve the empirical success of quantum mechanics or quantum information theory. Put differently, state assignments carry with them empirical consequences when they are quantum mechanical states. What needs to be reproduced for quantum communication to be successful is a system that behaves, in certain respects and degrees, like the system produced by the quantum information source. The upshot is that quantum information theory, under the constraint of preserving the empirical success of the theory, places constraints on the set of states associated with quantum messages, and correspondingly the types associated with those messages.

4.3 A Problem for the Definition of Quantum Information

Shannon's theory only countenances distinguishable types, but members of that set of types can be completely conventional. Quantum information theory countenances distinguishable and indistinguishable types, but these types cannot be completely conventional, else the empirical success of quantum mechanics would be compromised. This raises a worry. Quantum information theory is typically taken to be a generalization of Shannon information theory, as Bub (2007) has argued. That said, if there are different constraints on what kinds of systems can qualify as instances of Shannon information and quantum information, it raises the possibility that not every transfer of Shannon information is a transfer of quantum information, which would be counterintuitive, if not contradictory. In this subsection, I will argue that the solution to the problem is to recognize that quantum information theory involves conventional types much like Shannon's theory does.

Here's an example. Consider a qutrit, a three-dimensional quantum system, with associated orthogonal basis states $|a\rangle$, $|b\rangle$, and $|c\rangle$. Let us stipulate that the Shannon type of interest is one such that states $|a\rangle$ and $|b\rangle$, are tokens of the same type, call it type 0, and state $|c\rangle$ is a token of a different type, call it type 1. Now, suppose that an information source produced a sequence of systems in the states $|a\rangle|b\rangle|c\rangle$. According to the Shannon types of interest, this is a token of 001. Suppose the destination produces a sequence of system in the states $|a\rangle|a\rangle|c\rangle$, which is also a token of 001. If we take high entanglement fidelity to be the appropriate success criterion for quantum information transfer, then we have successful Shannon information transfer without successful quantum information transfer. The quantum tokens at the source and destination are orthogonal to one another. Put differently, the natural structural criterion of success in Shannon's theory, high entanglement fidelity, seems to be at odds with the view that quantum information theory is a generalization of Shannon information theory.

One might object and suggest that if a user identifies systems that would ordinarily be described using distinct quantum states, then those states ought to be represented by a single state rather than two. In the example examined here, just treat the system as though it were in two quantum states, perhaps $|c\rangle$ and $|d\rangle$ corresponding to 1 and 0, respectively. Thus, the systems would be described as being in the states $|d\rangle|d\rangle|c\rangle$ corresponding to 001. We assign arbitrary states to conventional types in Shannon's theory; why not do the same with quantum information theory? If we coarse-grain the representation of the system in this way, then there will be successful transfer of both Shannon and quantum information.

Set aside the problem of engaging in a deliberate act of ignorance. I certainly do not deny that one can misrepresent qutrits using a two-dimensional Hilbert space, but that misrepresentation has consequences. It has the consequence of undermining the predictive and explanatory power of quantum information theory. The reason why is that quantum states, dynamics, and observables are a tight package. If the quantum states are so and so, and the dynamics are so and so, then these are the measurement statistics we ought to observe. The trick is that if we treat qutrits like qubits, the tight connection between quantum states, dynamics, and observables will be compromised. The systems that are the outputs of the information source in question will behave like qutrits when one submits them to the corresponding measurements. Furthermore, one can discover that a sequence of N such systems can be used to distinguish between 3^N messages rather than the 2^N that one would expect from the state assignments that were given. The state assignments made render such possibilities unexpected and unexplained.

There is a way to preserve the claim that quantum information theory is a generalization of Shannon's theory, that all Shannon information transfer is quantum information transfer, and the definition of information provided earlier. The key is to recognize that quantum information theory will allow for more convention than one might expect, given the natural structural criterion for success, entanglement fidelity. Entanglement fidelity is the most discriminating means of individuating quantum types. That said, a user of a quantum communication system is not required to be interested in quantum information transfer protocols that preserve entanglement. One might simply be interested in preserving the behavior of individual systems emitted from a quantum information source. If this is what one is interested in, then one should use local fidelity of a process to discriminate quantum types, as described in Barnum and colleagues (1996). Continuing this line of reasoning, one might only be interested in preserving the behavior associated with some degrees of freedom, but not others, as in the qutrit example introduced previously in this chapter. So, users of a quantum communication system are free to stipulate which quantum types are required to be reproducible at the destination for communication to be a success, but these types need not be as fine-grained as those individuated by entanglement fidelity. In fact, a single type can have orthogonal tokens.

In Shannon's theory, tokens of a conventionally defined type *can always be assigned the same state* because of the lack of structure on Shannon's theory. Aside from the constraint that states are assigned only to distinguishable systems, a user is free to locate systems in Shannon's theory as they please. Quantum information theory has far more structure, and under the constraint of preserving the success of the theory, tokens of conventionally determined quantum types *cannot always be assigned the same state*. In the example of the qutrit that was used as a two-dimensional communication system is the case in point. A quantum type of interest was identified disjunctively, either $|a\rangle$ or $|b\rangle$. So, it is important to differentiate the constraints that users of a theory face when locating systems in the space of possibilities the theory countenances, assigning *states* to systems, and the constraints one faces for choosing a criteria for successful information transfer, a specification of relevant *types*. They are not generally the same.

Quantum information theory places constraints on possible quantum types. Quantum types can be bifurcated into different classes. The most basic and finest-grained quantum types are just those corresponding to density operators on a Hilbert space, tokens of which behave in just the way quantum mechanics suggests that they should. The other class of quantum types is going to be that composed by grouping members of the basic class of quantum types, as in the qutrit example, but also in cases of processes where local fidelity is preserved, or even entanglement fidelity, assuming that one doesn't demand that fidelity be maximized in a communication protocol, but allows for some deviation in relevant behavior of the system produced by the source and the one produced by the destination. Now, if a system is not successfully represented by a density operator on a Hilbert space, then it just cannot qualify as a token of *any* type of quantum information. A good example would be PR-boxes associated with Popescu and Rohrlich (1994), which exhibit correlations that are not representable by density operators on a Hilbert space. So, users have tremendous flexibility in defining quantum types that coincide with the representational power of quantum theory, but they cannot define *quantum* types that exceed that power.

5 Conclusions

Van Fraassen has importantly drawn our attention to the indexical aspect of scientific representation and Ruetsche has directed us to the appropriate way to interpret scientific theories. The Ruetsche edict has placed constraints on how one self-locates in theoretical space. One had better do so in a way that preserves the success of a theory. Thinking about self-location under that constraint taught us something important about information generally, and about Shannon information and quantum information and their differences in particular.

Consider the general definition of information again:

Information$_t$ is what is produced by an information$_t$ source that is required to be reproducible at the destination if the transmission is to be counted as a success.

This definition functions perfectly well for Shannon information theory and quantum information theory when it is specialized to them. What we have learned is that the definition has a striking conventional element to it, despite appearances to the contrary. Users of an information theory will be able to fix the types of interest for purposes of communication, types that might be defined conventionally, and in doing so fix the appropriate success criteria for communication.

In order for Shannon's theory to successfully determine the minimal resources necessary and sufficient for communication, different types of Shannon information need to be distinguishable from one another. That said, they need share no other properties except those determined by convention alone for successful information transfer. Subject to the initial stipulation, *any* system can be located in Shannon's theory as *any* message. So, same types can always be assigned the same state.

Unlike Shannon's theory, quantum information theory places fewer restrictions on the types it countenances insofar as it countenances distinguishable and indistinguishable types. That said, in order to preserve the success of quantum information theory, there are more constraints on how one locates systems in the theory. It is not the case that tokens of a legitimate quantum type can always be assigned the same state. Legitimate quantum types can be constructed by grouping members of the types (successfully) associated with density operators on a Hilbert space. This is what allows us to consider any instance of Shannon information an instance of quantum information transfer, in keeping with the view that quantum information theory is a generalization of Shannon information theory.

What have these considerations told us about the general definition of information? Well, that it functions perfectly well. It helps us understand what information is, and highlights differences between information theories. The definition should function appropriately for technical theories of information associated with compression. The definition is impressively general. Specializations of it to various theories will illustrate the differences in different types of information. We have seen this with Shannon's theory and with quantum information theory. One might develop a theory of, say, genetic information, on its basis. In doing so, the definition will serve to highlight the constraints that the theory of genetics would place on types of genetic information. For those interested in such theories, there is no better place to start.

Acknowledgments

This chapter arose through an invitation to speak at the conference *What Is Quantum Information?* The chapter benefited enormously from conversations with participants at the conference. I also thank Soazig Le Bihan for her comments on drafts.

References

Barnum, H., Caves, C., Fuchs, C., Jozsa, R., and Schumacher, B. (1996). "Noncommuting Mixed States Cannot Be Broadcast." *Physical Review Letters,* **76**: 2818–2821.

Bub, J. (2007). "Quantum Information and Computation." Pp. 555–660 in J. Butterfield and J. Earman (eds.), *Philosophy of Physics. Part B.* Amsterdam: Elsevier.

Caves, C., Fuchs, C., and Schack, R. (2007). "Subjective Probability and Quantum Certainty." *Studies in History and Philosophy of Modern Physics,* **38**: 255–274.

Duwell, A. (2008). "Quantum Information Exists." *Studies in History and Philosophy of Modern Physics,* **39**: 195–216.

Goodman, N. (1968). *Languages of Art.* Indianapolis, IN: Hackett Publishing Company.

Hughes, R. I. G. (1997). "Models and Representation." *Philosophy of Science* **64**: S325–S336.

Popescu, S. and Rohrlich, D. (1994). "Quantum Nonlocality as an Axiom." *Foundations of Physics,* **24**: 379–385.

Ruetsche, L. (2011). *Interpreting Quantum Theories: The Art of the Possible.* Oxford: Oxford University Press.

Schumacher, B. (1995). "Quantum Coding." *Physical Review A,* **51**: 2738–2747.

Shannon, C. (1948). "The Mathematical Theory of Communication." *Bell System Technical Journal,* **27**: 379–423, 623–656.

Suárez, M. (2004). "An Inferential Conception of Representation." *Philosophy of Science,* **71**: 767–779.

Timpson, C. G. (2004). *Quantum Information Theory and the Foundations of Quantum Mechanics.* PhD thesis, University of Oxford.

Timpson, C. G. (2008). "Quantum Bayesianism: A Study." *Studies in History and Philosophy of Modern Physics,* **39**: 579–609.

Timpson, C. G. (2013). *Quantum Information Theory and the Foundations of Quantum Mechanics.* Oxford: Oxford University Press.

Van Fraassen, B. (2008). *Scientific Representation: Paradoxes of Perspective.* Oxford: Oxford University Press.

3

Information, Communication, and Manipulability

OLIMPIA LOMBARDI AND CRISTIAN LÓPEZ

1 Introduction

The roots of the concept of information can be traced back to antiquity, in particular, to Aristotelian *hylemorphism*, that is, the doctrine according to which individuals (primary substances, *prôtai ousiai*) are composed of matter (*hyle*) and form (*eidos*): whereas matter is mutable and corruptible, form is what persists and makes something what it is. As opposed to mere opinion, knowledge is the process of apprehending the *form* of the objects. This idea passed to medieval scholasticism: in individual objects, matter is *informed* (*informatio materiae*); knowing amounts to abstracting the form from the object.

According to Adriaans (2013), the term 'information' almost vanished from the philosophical discourse in modern times, but gained popularity in everyday language. The link between information and knowledge, as well as the process view of information, survived for a long time in colloquial discourse. For instance, in Daniel Defoe's novel, Robinson Crusoe refers to the education of Friday as his "information." However, since the nineteenth century, the concept of information has gradually become detached from the process view, and it has acquired the status of an abstract mass noun: a text is endowed with the capacity to inform because it "contains" information. This idea entered science: information turned out to be something that can be stored and measured (Fisher 1925).

With World War II, the strategic value of information was immediately acknowledged. But during wartime, information was important not only because it can be stored and measured, but mainly because it can be transmitted; in particular, it can be securely and secretly transmitted: coding became a priority. In postwar times, the link between information and communication grew stronger with the explosion of communication technologies: telephony, radio and television broadcasting, and the Internet. Now the interest is not only transmitting information, but also transmitting it efficiently. Thus, "information theory" has become the core of the training of communication engineers, for whom the main problem consists in optimizing the transmission of information by means of physical carriers whose energy and bandwidth are constrained by technological and economic limitations.

In this communicational context, the classical locus is the paper where Claude Shannon (1948) introduces a precise formalism designed to solve certain specific technological problems in communication engineering (see also Shannon and Weaver 1949). The very existence of such formalism might lead us to suppose that there are no problems about how to understand the concept of information. Nevertheless, this is not the case: even on the basis of a same formal substratum, different interpretations of the concept of information still coexist (see Lombardi 2004; Lombardi, Fortin, and Vanni 2015). The situation becomes more complex with the advent of quantum coding and entanglement-assisted communication (Schumacher 1995), which combines the difficulties in the understanding of the concept of information with the well-known foundational puzzles derived from quantum mechanics itself.

The purpose of this chapter is not to address the question about the nature of information. Perhaps there is no single concept that can subsume all the ideas surrounding the term 'information': knowledge, capacity to be stored, coding, transmission. Rather, the aim is to supply criteria to identify information in a communicational context. For this purpose, in Section 2 we will begin by recalling how communication is introduced in Shannon's formalism, and by considering the different interpretations of the formalism that can be implicitly or explicitly found in the literature. Then, in Section 3 we will consider the additional challenges raised by the advent of entanglement-assisted communication, and some conceptual responses of those challenges. The complexity of the situation described in the previous sections will lead us to come back, in Section 4, to the basic idea of communication and its essential notes; this will allow us to analyze how the different views about information would account for communication and to conclude that none of them supplies adequate criteria to identify information in the communicational context. In Section 5, we will argue that the communication of information is a process that, in a certain sense, involves the notion of causation and the asymmetry implicit in it; therefore, we will analyze, among the different concepts of causation, which can serve to apply to our case of communication. Finally, Section 6 will be devoted to support the claim that the manipulability accounts of causation supply the philosophical tools to characterize the transmission of information in a communicational context. This approach does not intend to have elucidated the very nature of information or to have given the right definition of the concept. Nevertheless, many conundrums around the concept of information, when applied to communication, are solved or simply vanish in the light of this manipulability view of information.

2 One Formalism, Different Interpretations

In his paper "The Mathematical Theory of Communication" (1948), Shannon offered precise results about the resources needed for optimal coding and for error-free communication. However, in the present context, it is sufficient to recall three parts of the communication system as Shannon describes it (see Figure 3.1):

- A source S, which produces the information $H(S)$ to be transmitted. S is a system with a range of possible states s_i, each one with its own probability $p(s_i)$.
- A destination D, which receives the information $H(D)$. D is a system with a range of possible states d_j, each one with its own probability $p(d_j)$.
- The correlations between the states of the source S and the destination D, expressed by the conditional probabilities $p(d_j/s_i)$.

With these elements it is possible to define the *Shannon entropy* of the source $H(S)$ and of the destination $H(D)$, usually interpreted as the average amounts of information generated at the source and received at the destination, respectively,

$$H(S) = -\sum_{i=1}^{n} p(s_i) \log p(s_i) \qquad H(D) = -\sum_{j=1}^{m} p(d_j) \log p(d_j) \qquad (1)$$

and the so-called '*transinformation*' or '*mutual information*' $H(S; D)$, conceived as the average amount of information produced at the source and received at the destination,

$$H(S; D) = -\sum_{i=1}^{n} p(s_i, d_j) \log p(s_i, d_j) \qquad (2)$$

where $p(s_i, d_j) = p(s_i)p(d_j/s_i)$ is the joint probability of s_i and d_j.

Although the ideal of communication is that all the information generated at the source, and it alone, arrives to the destination, in real situations this is not the case: *equivocity E* is the average amount of information generated at S but not received at D, and *noise N* is the average amount of information received at D but not generated

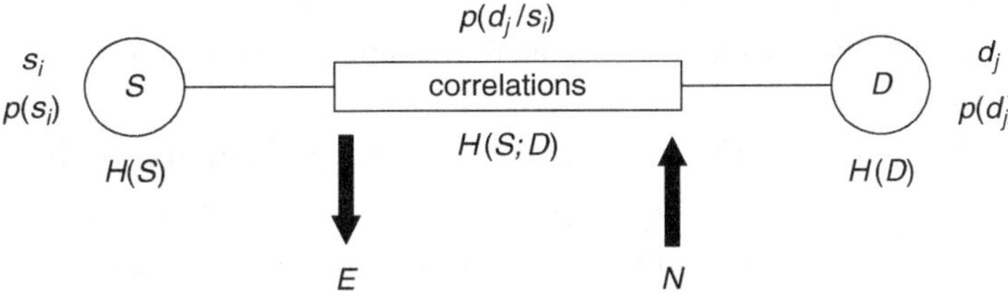

Figure 3.1. Sketch of a communication system.

at S. Therefore, independently of the way in which the amounts of information are computed, the following relations hold (see Figure 3.1):

$$H(D) = H(S) - E + N \quad H(S;D) = H(S) - E = H(D) - N \tag{3}$$

In real communications, the goal is to reduce noise and equivocation to a minimum by improving the features of the channel to avoid information loss, and by including filters to block noise. On the basis of these magnitudes, four kinds of situations can be distinguished:

- Deterministic: the values of E and N are minimum ($E = N = 0$), and the value of $H(S;D)$ is maximum ($H(S;D) = H(S) = H(D)$). There is a perfect correlation between the states of the source and the states of the destination.
- With noise but without equivocity: only the value of E is zero ($E = 0$ and $N \neq 0$). Therefore, the value of $H(S;D)$ is not maximum ($H(S;D) = H(S) = H(D) - N$). There are states of the source that are correlated to more than one state of the destination.
- With equivocity but without noise: only the value of N is zero ($N = 0$ and $E \neq 0$). Therefore, the value of $H(S;D)$ is not maximum ($H(S;D) = H(S) - E = H(D)$). There are states of the destination that are correlated to more than one state of the source.
- With equivocity and with noise: neither the value of E nor the value of N are zero ($N \neq 0$ and $E \neq 0$). Therefore, the value of $H(S;D)$ is not maximum ($H(S;D) = H(S) - E = H(D) - N$). There are states of the destination correlated to more than one state of the source and vice versa.

In general, the information produced at the source is encoded before entering the channel that transmits the information, and decoded after leaving the channel and before being received at the destination. Claude Shannon (1948) and Benjamin Schumacher (1995) demonstrated theorems that supply the optimal coding in the classical and quantum case, respectively. Nevertheless, this aspect will be not central in the following discussion (for an analysis of Shannon's formalism, see Lombardi, Holik, and Vanni 2016b).

Up to this point, the formalism says nothing about the nature of information: the magnitudes represented by $H(S)$, $H(D)$, $H(S;D)$, E, and N have not been yet endowed with an interpretation. It is at this point where different views begin to diverge.

Many authors, especially those coming from philosophy or from cognitive sciences, take the most traditional perspective, which establishes a close link between information and knowledge. We receive information when we read a newspaper or watch the news on TV. Students acquire new information at school.

When somebody hides information, he doesn't want us to know something we should know. From this view, information is something that modifies the state of knowledge of those who receive it.

Although the link between information and knowledge is strongly present in colloquial language, it is not a feature exclusive of the everyday notion of information: this *epistemic* view is very commonly found in the philosophical and scientific literatures. For example, it has been applied in psychology and cognitive sciences to model human behavior as an information-flow process (MacKay 1956) and to conceptualize the human abilities of acquiring knowledge (Hoel, Albantakis, and Tononi 2013). Even in physics it can be read that we get "knowledge, i.e., information, of an object only through observation" (Zeilinger 1999: 633) and, as a consequence, one can talk about the "measure of information or knowledge" (Brukner and Zeilinger 2009: 681–682).

In the philosophical field, the epistemic view of information is implicit in authors interested in different aspects of information (e.g., MacKay 1969; Dunn 2001). In particular, it can be recognized in some semantic approaches to information. For instance, Fred Dretske's explicit purpose is to formulate a *semantic* theory of information capable of grasping what he considers the nuclear sense of the term 'information': "A state of affairs contains information about X to just that extent to which a suitable placed observer could learn something about X by consulting it" (1981: 45). According to him, the states of the Shannon source of information may be endowed with "informational content," which is expressed in the form "S is F," where the letter S is an indexical or demonstrative element referring to some item at the source. For Dretske, the statement "a state d of the destination carries the information that S is F" amounts to the statement "the conditional probability of S's being F, given d (and k), is 1 (but, given k alone, less than 1)," where k stands for what the destination already knows about the possibilities existing at the source (1981: 65). This might lead us to suppose that semantics is always involved when the concept of information is linked to that of knowledge: what is learned in the acquisition of information would be something with semantic content. However, this is not the case. When a physicist conceives information as something that supplies knowledge, he is not thinking in a generic content of the form "S is F," but in knowing which was the state occurred at the source by having access only to the destination. In other words, from an epistemic standpoint, it can be considered that a state d_j of the destination carries the information about the occurrence of the state s_i of the source if one knows that s_i occurred by knowing the occurrence of d_j, and this happens when $p(s_i) > 0$ (i.e., the occurrence of s_i is not certain, since in that case such an occurrence would not contain information) and $p(s_i/d_j) = 1$ (Lombardi 2005). In this case, although information supplies knowledge, it has no other content than that of identifying what state occurred at the source from a range of possible states.

From this epistemic perspective, a physical connection between the source and the destination is not required for the transmission of information. Nevertheless, the probabilities that link the states of the source with those of the destination cannot result from mere accidental correlations. In fact, when the correlation between two variables is merely accidental, the value of one of them says nothing about the value of the other. For instance, even if the properties P and Q are perfectly correlated – i.e., every P is Q and every Q is P – this does not guarantee that we can know that "x is Q" by knowing that "x is P": if the correlation between P and Q is a mere coincidence, x's being P tells us nothing about x's being Q. In other words, even the exceptionless one-to-one correlation, if accidental, does not supply knowledge. Therefore, the epistemic view requires that the probabilities involved in the communicational context be not de facto correlations, but manifestations of underlying lawful regularities. Indeed, commonly there is an elaborate body of theory that stands behind the attribution of probabilities (Lombardi 2004).

From a completely different viewpoint, in physics and communication engineering, the concept of information is detached from the notion of knowledge: information is endowed with a physical meaning. In these fields, the interest focuses on the physical resources needed for the transmission of information. One of the first applications of Shannon's theory was radio broadcasting: certainly, one can learn many things when listening to the radio, but one can also enjoy good music in an exclusively aesthetic experience, with absolutely no epistemic aim. Communication can also be used only with control purposes: for instance, the source might be operated by a surgeon, who performs a surgical procedure by means of the remote manipulation of the destination device.

For this physical view, information is always embodied in a physical substratum. As expressed by the well-known slogan "no information without representation," the transmission of information between two points of the physical space necessarily requires an information-bearing signal, that is, a physical process propagating from one point to the other. Rolf Landauer expresses clearly this assumption when he claims that "[i]nformation is not a disembodied abstract entity; it is always tied to a physical representation" (1996: 188).

The need of a physical representation of information says nothing yet about the nature of information: it could be conceived as a kind of non-physical emergent of particular physical processes. However, frequently this view is combined with the idea that information is a physical magnitude, something that can be generated at one point of the physical space and transmitted to another point; it can also be accumulated, stored, and converted from one form to another. Perhaps the best-known defender of this view is again Landauer, who claims that "information is physical" (1991, see also Landauer 1996; Rovelli 1996). Some authors conceive information as a physical entity with the same ontological status as energy, which can also be transmitted, accumulated, and converted from one form to another (Stonier 1990, 1996). More cautiously, Carlo Rovelli (personal communication) stresses the analogy between the historical emergence of the concept of information and that of the concept of energy. From beginning as a concept designed to measure the capacity of a system to perform a certain task – work – the concept of energy was gradually losing its subsidiary role and became one of the fundamental concepts of physics, perhaps the concept that currently plays the most important role in the unification of the different domains of physics. In the light of the strong presence of the concept of information in the different areas of contemporary physics, it is reasonable to wonder whether it is following a historical path analogous to that energy took in the past, acquiring an independent and unifying physical content.

These different views about the concept of information have been implicitly present in the literature since the formulation of Shannon's theory, showing that the agreement about a certain formalism is not sufficient to lead to agreement about its content. Nevertheless, the challenges posed by the nature of information acquired a renewed relevance with the advent of entanglement-assisted communication.

3 A New Challenge: Entanglement-Assisted Communication

During the past decades, so-called quantum information theory has become an ever growing field, devoted to exploit quantum resources to perform informational tasks. In particular, entanglement has been used as a highly important resource in communication. Teleportation is one of the paradigmatic examples of entanglement-assisted communication.

Although a direct result of quantum mechanics, teleportation appears as a weird phenomenon when described as a process of transmission of information. Broadly speaking, an unknown quantum state $|\chi\rangle$ is transferred from Alice to Bob with the assistance of a shared pair of particles prepared in an entangled state and of two classical bits sent from Alice to Bob (the description of the protocol can be found in any textbook on the matter; see, e.g., Nielsen and Chuang 2010). In general, the perplexity this situation raises is explained by saying that a very large (strictly infinite) amount of information is transferred from Alice to Bob by sending only two bits, and without any physical link other than the classical channel through which the classical bits are transmitted.

In his detailed analysis of teleportation, Christopher Timpson poses the two central questions of the debate: "First, how is so much information transported? And second, most pressingly, just how does the information get from Alice to Bob?" (2006: 596).

Regarding the first question, it is usually said that the amount of information generated at the source is, in principle, infinite because two real numbers are necessary to specify the state $|\chi\rangle$ among the infinite states of the Hilbert space. Even in the case that a coarse-graining is introduced in the Hilbert space, the amount of information is immensely greater than the two bits sent through the classical channel. Nevertheless, these supposed difficulties might be dissolved by distinguishing the stage of the generation of information and the stage of the encoding of information. On this basis, teleportation is not a process of transmission of information between a source and a destination, but a part of the complete arrangement for communication, in particular, a way, among others, in which the quantum state that encodes the information generated at the source can be *transposed* between transmitter and receiver: "'quantum teleportation' [...] is a rather exotic example of a transposition process" (Schumacher 1995: 2741). In other words, teleportation is a physical process that allows a quantum state to be transferred between two spatially separated quantum systems without leaving a copy behind, and this process does not need to be conceptualized in informational terms to be understood: it is only a part of a protocol of transmission of information (for a full argument, see Lombardi, Holik, and Vanni 2016a).

However, even from this reading of teleportation, the second question preserves its meaning: how does information get from Alice to Bob? This question points to the very interpretation of the concept. Whereas in traditional communication the information is always transferred by some physical carrier, in the case of teleportation there is no physical signal that carries the teleported state from the transmitter to the receiver: the quantum channel is constituted by the entanglement between the quantum particles of the pair. As a consequence, some physicists proposed other physical links to play the role of carriers of information. For instance, according to David Deutsch and Patrick Hayden (2000), the quantum information travels hidden in the classical systems that Alice sends to Bob, which carry the two bits necessary to complete the protocol. For Richard Jozsa (1998) and Roger Penrose (1998), in turn, the information travels backward in time, from the event at which Alice performs her operation on her particle to the event at which the entangled pair was produced, and then travels forward to the future, until the event at which Bob performs his operation on his particle.

These accounts of the way in which information is transferred in teleportation are implicitly tied to a physical interpretation of information: they insist on the search for a physical link, that is, for a signal that carries the information from Alice to Bob. Jozsa and Penrose think of a quantum channel, but since there is no direct quantum signal between Alice and Bob, then the channel extends toward the past up to the production of the entangled pair, and then toward the future. For Deutsch and Hayden, by contrast, the two classical systems that travel from Alice to Bob are the physical carriers where the information encoded in the quantum state is hidden during the transmission. Although both views answer the question "how does information get from Alice to Bob?" they sound rather bizarre to the ears of most physicists, used to forward causation and physically manifested signals.

The difficulties of the physical view of information when facing teleportation do not have to lead us to suppose that the epistemic interpretation fares better. In the case of quantum coding, the states s_i of the source are encoded by means of states $|a_i\rangle$ belonging to a Hilbert space $\mathcal{H}_M$ of dimension $\dim(\mathcal{H}_M) = d$, which may be non-orthogonal, and whose respective probabilities of occurrence are $p(a_i) = p(s_i)$. The mixture of the states $|a_i\rangle$ can be represented by the density operator

$$\rho = \sum_{i=1}^{n} p(a_i)|a_i\rangle\langle a_i| \in \mathcal{H}_M \otimes \mathcal{H}_M \tag{4}$$

whose von Neumann entropy is

$$S(\rho) = -\mathrm{Tr}(\rho \log \rho) \tag{5}$$

Only in the case that the $|a_i\rangle$ are mutually orthogonal, the von Neumann entropy is equal to the Shannon entropy, $S(\rho) = H(A)$, but in the general case,

$$S(\rho) \leq H(A) \tag{6}$$

Moreover, the so-called *Holevo bound* (Holevo 1973) establishes an upper bound for the mutual information in the case of quantum coding. When the coding states are pure states $|a_i\rangle$, the bound is given by:

$$H(A;B) \leq S(\rho) \tag{7}$$

As a consequence, from (6) and (7) it is clear that, in the case of quantum coding by means of non-orthogonal states,

$$H(A;B) \leq H(A) \tag{8}$$

This result expresses the fact that, when the quantum states are non-orthogonal, it is not possible to distinguish them from each other by measurement. In his paper on quantum coding, Schumacher (1995: 2739) characterizes the *accessible* informa-tion as the maximum amount of information that can be recovered by measure-ment, which is given by $H(A;B)$. Therefore, when messages are encoded by means of non-orthogonal – and then non perfectly distinguishable – quantum states, the accessible information is lower than the information produced at the source: there is *always* a loss of information between the message source and the message destination measured by an equivocity $E \neq 0$ computed as $H(A) - H(A;B)$. These results show that, when generic quantum states are used for coding, the information produced at the source cannot even in principle be recovered at the destination: the empirical indistinguishability of the coding states makes it impossible to know what happened at the source by observing the destination. This means that, from an epistemic view of information, a paradigmatic example of entanglement-assisted communication such as teleportation would be *in principle* a defective communication process, which should be improved by means of redundancy, that is, by sending the same information many times.

In short, both the epistemic and the physical views of information face conceptual difficulties in the case of teleportation. A way out of the conundrum is that followed by Christopher Timpson, who cuts the Gordian knot of teleportation by adopting a deflationary approach to information. According to this approach, 'information' is an abstract noun and hence does not refer to a spatiotemporal particular, to an entity or a substance (Timpson 2004, 2013; see also Duwell 2008). As a consequence, "there is not a question of information being a substance or entity that is transported, nor of 'the information' being a referring term" (Timpson 2006: 599). From this viewpoint, asking how information gets from Alice to Bob makes no sense: the only meaningful issue in teleportation is about the physical processes involved in the protocol.

The deflationary approach was originally based on the type/token distinction: when information is transmitted, the aim is to reproduce at the destination another token of the same type as that produced at the source (Timpson 2004, 2006; for criticism, see Lombardi, Fortin, and López 2016). In more recent works, the distinction is generalized in terms of sameness of pattern or structure: "the distinction may be generalized. The basic idea is of a pattern or structure: something which can be repeatedly realized in different instances" (Timpson 2013: 18). The idea of isomorphism suggests that, in communication, the states of the source and the states of the destination must be linked through a one-to-one mapping: "the success criterion is given by an arbitrary one-to-one mapping from the set of the letters of the source to the set of the letters of the destination" (Duwell 2008: 200). However, as we have seen in Section 2, this is not always the case: the one-to-one correlation holds in deterministic situations, but not in generic cases with equivocity and/or noise. But even worse is the fact that isomorphism says very little about information: mere correlation, even if perfect, does not guarantee communication.

The discussions of the present and the previous sections show the subtleties surrounding the concept of information, even in the framework of a single formalism. This situation shows the convenience of returning to the basics of the matter to elucidate the very idea of communication.

4 What Is Communication?

In general, the first distinction introduced in the debates about the concept of information is that between a semantic and a statistical approach to information. The first approach conceives information as something that carries semantic content (Bar-Hillel and Carnap 1953; Bar-Hillel 1964; Floridi 2011) and, as a consequence, is related to semantic notions such as meaning, reference, and truth. According to the statistical approach, information is concerned with the compressibility properties of sequences of states of a system and/or the correlations between the states of two systems, independently of any meaning or reference. In the present chapter, the discussion was confined to the domain of statistical information, although this does not imply disqualifying the attempts to add a semantic dimension to a statistical formalism (MacKay 1969; Nauta 1972; Dretske 1981).

However, the focus on the statistical approach is not yet sufficiently specific. In fact, in the domain of statistical information there are different contexts in which the concept of information is defined. In the traditional communicational context, information is primarily something to be transmitted between two points for communication purposes; it is in this context that Shannon's paper became a classic. In the computational context, by contrast, information is something that has to be computed and stored in an efficient way; in this context, algorithmic complexity measures the minimum resources needed to effectively reconstruct an individual message (Solomonoff 1964; Kolmogorov 1965, 1968; Chaitin 1966). In this chapter, we were only dealing with the concept of information in the communicational context.

Many different definitions of the concept of communication can be found in the literature, most of them involving semantic notions such as meaning, epistemic notions such as understanding, and even notions referring to feelings and to emotions. Here we are not interested in giving a definition of communication, but only in isolating some of its essential notes, in particular those related to Shannon's formalism. The aim is to identify in which situations it can be said that there is communication and, consequently, transmission of information.

Independently of how the measure of information is mathematically defined, there are certain minimum elements that can be abstracted to characterize a communicational context. From a very general perspective, communication requires a source S with its different states, which produces the information to be transmitted, a destination D also with its states, which receives the information, and a channel through which information is transmitted from the source to the destination. In the context of this abstract framework, communication requires that a certain action performed at the source modifies the destination so as to establish a correlation between the state of the source and the state of the destination. Let us notice that this characterization does not involve knowledge, not even in the weak sense of a subject that identifies what state occurred at the source from knowing the state occurred at the destination: as stressed in Section 2, the state of the destination can be manipulated from the source for exclusively control purposes. Moreover, no perfect correlation is required: as explained also in Section 2, non-perfect correlation, manifested as equivocity and/or noise, can be corrected by means of redundancy and/or filters that preserve communication.

This characterization, although very abstract, includes two essential notes of communication, which are not independent but linked to each other:

- *Asymmetry.* Communication is an asymmetric process: the source sends information and the destination receives it. Although in a following stage the roles can be interchanged, in each run of the communication process source and destination are clearly different and cannot be confused.
- *Production.* What happens at the source modifies what happens at the destination, producing a specific modification of the state of the destination. In other words, the asymmetry is not a merely formal relationship, but a physical connection that links events occurred in different space-time locations.

These two features are not manifested in Shannon's formalism taken at face value, without adding explanations related to producing, sending, and receiving information. For instance, when information theory is presented from an exclusively formal perspective as a chapter of the theory of probability (see Khinchin 1957; Reza 1961), its basic concepts are defined in terms of random variables and probability distributions over their possible values (see Cover and Thomas 1991), with no mention of sources, destinations, signals, and transmission. In turn, the essential notes of communication cannot be obtained from a deflationary view of information (Duwell 2008; Timpson 2013): asymmetry and production find no place in the context of a view that conceives the link between source and destination as merely sameness of form. Therefore, we are still left with the traditional epistemic and physical interpretations of information to deal with communication.

The following step consists in applying this abstract characterization of communication to various situations in order to decide whether they are really communicational situations. The purpose will be to use those situations as a probe to test the different views of information in the light of the generic idea of communication.

The first case is that of a contingent agreement. Let us suppose that during the past month, the results of the Mexican national lottery agree with the results of the Australian national lottery. This case cannot be conceived as a communication situation since it lacks both asymmetry and production: nothing can be done in one place to modify the state in the other place. However, from an exclusively formal viewpoint, a correlation between the two sequences of events can be defined. But in this case (i) the correlations are de facto, with no nomological link between them, and (ii) there is no physical link between the two ends. This means that both the epistemic and the physical interpretations correctly disqualify this case as a communicational situation.

The second case is that of a TV transmitter T that broadcasts an electromagnetic signal, which is received by two television sets TV_A and TV_B. Again, from an exclusively formal viewpoint, a correlation between the states of the two TV sets can be defined. But is there communication between them? As in the previous case, this one cannot be conceived as a communication situation: it is completely symmetrical, and the modification of the state of a TV set cannot be produced by changing the state of the other. However, in this case the two interpretations do not agree. In fact, (i) there is no physical interaction between the two TV sets, but (ii) the correlations between them are not accidental but the result of the physical dependence of the states of TV_A and TV_B on the states of the transmitter T. Therefore, whereas the physical interpretation will correctly conclude that this is not a communication situation, the epistemic interpretation will qualify it as communicational to the extent that one can learn what is the state of one of the TV sets by consulting the other (see Dretske 1981: 38–39).

The third case is an EPR experiment, where Alice and Bob perform a measurement on their own particle of a pair in an entangled pair (Einstein, Podolsky, and Rosen 1935). Once again, a formal correlation between the results of Alice's and Bob's measurements can be defined. However, as in the previous cases, this one lacks both the asymmetry and the production necessary to be conceived as a communication situation: Alice can do nothing to produce a specific modification of the result of Bob's measurement, and vice versa. Nevertheless, since there is a nomological link underlying the correlations, from an epistemic interpretation, the situation might qualify as communicational. This contrasts with the usual view of the EPR experiments, implicitly based on a physical interpretation of information, according to which there is no transmission of information between the two particles, because the propagation of a superluminal signal from one particle to the other is impossible: there is no information-bearing signal that can be modified at one point of space in order to carry information to the other spatially separated point. This means that only the physical view correctly classifies the situation as non-communicational.

The second and the third situations act as a silver bullet for the epistemic view, since they make clear that merely epistemic relations are not sufficient for communication. In other words, the epistemic interpretation of information, although based on a natural conception of information as linked to knowledge, loses its reference to the communicational context in which the information is defined. Therefore, up to this point the physical interpretation seems a better option. However, the seeming advantage disappears when entanglement-assisted communication is considered. As explained in the previous section for the case of teleportation, a communication protocol can be implemented without a physical signal between the source and the destination of information. Therefore, the requirement of a physical carrier of the information that travels from two points of space turns out to be too demanding when the task is to identify situations of communication.

Perhaps a way out of this puzzle can be found in the concept of causation: independently of the nature of the channel, and the nature of the link between source and destination, communication requires that what happens at the source in a certain way produces, *causes*, what happens at the destination. Moreover, the causal connection between source and destination would recover the asymmetry of the communication process. Although this seems a promising strategy, the main challenge that it has to face is the elucidation of the very concept of causation.

5 What Causation? The Manipulability Account

Although causation may offer a way out of the problem of interpreting the concept of information, the risk is to elucidate an obscure concept – information – by means of another, even more obscure concept – causation. In fact, the concept of causation is one of the most discussed topics in the history of philosophy; the difficulties that surround it have led many authors to follow Humean ideas and to advocate for the "the complete extrusion" of the word 'cause' from physics (Russell 1912). Of course, we will not follow this path.

Despite the long tradition of the counterfactual approaches to causation, in a physical framework it seems reasonable to appeal to a physical conception of causation. From the physical perspective, causation has been conceived in terms of energy flow (Fair 1979; Castañeda 1984), of physical processes (Russell 1948; Dowe 1992), and of property transference (Ehring 1986; Kistler 1998). However, all these views involve physical signals or space-time connections and, as a consequence, they are not adequate to elucidate a concept of information that does not require a physical interaction between source and destination.

On the other hand, independently of any philosophical discussion, both in everyday life as in science, people act as if there were real causal links, without considering if there is a space-time connection between the cause and its effect. Regarding causes, anyone distinguishes the case of pain on a finger due to a hammer blow from the appearance of the paperboy when the sun rises.

Similarly, a chemist clearly distinguishes the causal action of a catalyst in increasing the rate of a reaction from the mere correlation between the melting point and the color of an element.

The *manipulability accounts of causation* intend to capture this intuitive distinction. Their basic idea is that it is possible to draw the distinction between cause–effect relationships and mere correlations by means of the notions of manipulation and control. Nancy Cartwright (1979) stresses this central feature of causation by asserting that causal relationships are needed to ground the distinction between effective and ineffective strategies: an effective strategy proceeds by intervening at a cause in order to obtain a desired outcome. In a similar vein, Thomas Cook and Donald Campbell hold: "The paradigmatic assertion in causal relationships is that manipulation of a cause will result in the manipulation of an effect" (1979: 36). In other words, only causal relationships, but not mere correlations, are exploitable by us in order to bring about a certain outcome (Frisch 2014).

There are different manipulability accounts of causation. According to the early versions, causal terms need to be reduced to non-causal terms, such as free agency. For instance, Georg von Wright states that

to think of a relation between events as causal is to think of it under the aspect of (possible) action. It is therefore true, but at the same time a little misleading to say that if p is a (sufficient) cause of q, then if I could produce p, I could bring about q. For that p is the cause of q.

(1971: 74)

A more recent version of a manipulability account of causation is presented by Huw Price (1991) and Peter Menzies and Price, who attempt to develop an "agency" theory of causation:

an event A is a cause of a distinct event B just in case bringing about the occurrence of A would be an effective means by which a free agent could bring about the occurrence of B.

(1993: 187)

The basic premise of this approach is that, from an early age, we have direct experience not merely of Humean succession of events, but mainly of our acting as agents. Therefore, the notion of causation has to be tied to our "personal experience of doing one thing and hence achieving another" (Menzies and Price 1993: 194).

These first manipulability versions received several criticisms. On one hand, they were charged with circularity: "doing" and "producing" are already causal notions, which, therefore, cannot be legitimately used to define the notion of causation. On the other hand, manipulability is an anthropocentric notion; then, the resulting concept of causation is not sufficiently general since linked to human manipulation power. For instance, this account would not be able to identify the relationship between the gravitational attraction of the moon and the motion of the tides on the earth as a causal relation. In other words, the agent-based account of causation is inapplicable to causal relationships in which the manipulation of the cause by human beings is not practically possible.

The interventionist version of the manipulability account of causation, developed by James Woodward (2003, 2007; see also Hausman and Woodward 1999; Pearl 2000), comes to solve those criticisms. Woodward notices that the accounts of causation common among non-philosophers, statisticians, and theorists of experimental design have no reductionist aspirations: causation is a primitive notion that cannot be reduced to simpler and more basic concepts (Woodward 2013). Moreover, the author points out that causal attributions were traditionally made even before the advent of the Galilean idea of natural law (Woodward 2003: 171). On this basis, the interventionist approach does not intend to define causation in terms of non-causal notions, but to delimitate the domain of causation by means of the possibility of control or manipulation: the response to interventions is used as a probe to know whether a certain relation is causal (Woodward 2003: 21). As Mathias Frisch puts it, "the results of interventions into a system are a guide to the causal structure exhibited by the system" (2014: 78).

Woodward's interventionist proposal "focuses on the purposes or goals behind our practices involving causal and explanatory claims; it is concerned with the underlying point of our practices" (2003: 7). Two elements are central in the proposal: the characterization of causal relationships as relating *variables*, and the notion of "*intervention*" as an action that produces a change in the value of a variable. The former makes it possible to identify with precision the relata of the relation under scrutiny, and to conceptualize the changes introduced by manipulation as changes in the values of the variables (2003: 39). In this way, Woodward intends to capture, in non-anthropocentric language, the idea of determining whether X causes Y by means of an ideal experimental manipulation of the value of X.

Informally, regarding the relationship between X and Y, an *intervention* is a causal and exogenous process that modifies the value of X in a specific way:

the intuitive idea is that an intervention on X with respect to Y changes the value of X in such a way that if any change occurs in Y, it occurs only as a result of change in the value of X and not from some other source.

(Woodward 2003: 14; for a detailed definition, see 2003: 98)

In this case, it can be said that the relationship between X and Y is a genuine instance of causation. Let us consider the following example: on May 22, 1960, a very strong earthquake destroyed the city of Valdivia, Chile, where the seismograph recorded 9.5 on its scale. Three two-value variables can be defined: E, D, and S, whose values 1 represent the occurrence of the earthquake, the destruction of Valdivia, and the reading of 9.5 in the seismograph, respectively. If an intervention on E changed its value from 1 to 0 (by inhibiting the occurrence of the earthquake) without changing other environmental variables, then D and S would also change their values from 1 to 0 (the destruction of Valdivia would not occur and the seismograph reading would not be of 9.5, respectively); therefore, the relationships between E and D and between E and S are causal. On the contrary, the value of D would not change at all if an intervention on S changed its value from 1 to 0: regrettably, the destruction of Valdivia could not have been prevented by an exogenous modification of the reading of the seismograph, and this is the indication that the relation between D and S is not causal, but a mere correlation due to a common cause.

The interventionist approach to causation explicitly faces the criticisms directed to the previous manipulability views (see Woodward 2013). On one hand, the circularity criticism does not apply. In deciding whether X is a cause of Y, we can use assumptions about the causal relationship between *other* pairs of variables: for instance, the causal relationship between the intervention I and the variable X. Of course, this would be unacceptably circular if the aim were to define causation by reducing the concept to non-causal notions. But since this is not the purpose of the interventionist approach, there is no vicious circularity: the causal assumptions are not about the very relationship whose causal nature we are examining. On the other hand, the interventionist approach also eludes the charge of anthropocentrism, because the concept of intervention should be understood without reference to human action. The consideration of *possible* interventions admits a counterfactual formulation, which makes sense of causal claims in situations where interventions do not occur and even in cases in which they are impossible in practice.

Besides the traditional charges of circularity and anthropocentrism, other criticisms have been directed toward the interventionist approach to causation (see Woodward 2013). One of them is related to the use of counterfactuals: since the truth conditions for counterfactuals can be explained in terms of laws, the appeal to interventions is not necessary (Hiddleston 2005). But this objection is not as serious as it seems: the interventionist approach is not a counterfactual view of causation because the counterfactuals are not applied to the very relationship whose causal nature is to be determined. In turn, Nancy Cartwright characterizes the interventionist approach as "operationalist": it admits a single criterion to test causation, and leads to "withhold[ing] the concept [of cause] from situations that seem the same in all other aspects relevant to its application just because our test cannot be applied in those situations" (2002: 422). Independently of how appropriate these criticisms are, they are directed toward an approach that intends to elucidate the very concept of causation. But our concerns about causation are more modest: we are not interested in supplying a characterization of causation applicable in every circumstance in which the causal talk makes sense. Our only aim here is to explore the possibility of appealing to interventionist causation to characterize the informational relation between source and destination in a communicational context.

6 Communication and Manipulability

Let us recall that the concept of causation entered the scene with the purpose of solving the problems of the traditional epistemic and physical interpretations of information. To this end, the manipulability conception of causation, in particular in its interventionist version, seems appropriate because it embodies the two essential notes of communication: asymmetry and production. Precisely, there is transmission of information between source and destination when there is a causal link between them, where causation is conceived in manipulability terms.

It is easy to see that this view of information gives adequate answers when applied to situations as those of the lotteries or the TV sets: it correctly denies communication in both cases, since there is no causal link between the two ends of the setup: no intervention on one end can be used to manipulate the other end. But the main challenge is to apply this approach to quantum mechanics, in particular, to teleportation, where there seems to be a communicational channel with no physical interaction.

The first obstacle to be overcome is related to the application of the interventionist version of causation to quantum mechanics, which Woodward explicitly impugned:

The notion of an intervention with respect to one of the measurement events is not well defined in the EPR phenomena, because the distinction between intervening with respect to X and acting directly on both X and Y cannot be drawn.

(Hausman and Woodward 1999: 566; see also 2004)

Here the issue concerns the ontological interpretation of entangled systems in EPR-type experiments: Daniel Hausman and Woodward presuppose ontological holism, according to which the two particles of the entangled pair constitute an indivisible composite system: "In some way that is difficult to understand, the two particles constitute a single composite object, even though they may be at spacelike separation from each other" (Hausman and Woodward 1999: 566). It is quite clear that, from this ontological perspective, there cannot be an intervention on, say, the spin value a of particle A without intervening on the spin value b of particle B, to the extent that the two particles are parts of a single whole. In his discussion with the authors, Mauricio Suárez proposes to distinguish between the non-separable state of the entangled particles and the non-separability of the events involved in the experiment. On this basis, he concludes that "whether interventions are available in EPR (and quantum mechanics in general) is a complex and contextual question that does not have a unique or uniform answer" and that "different combinations of causal hypotheses under test and interpretations of quantum mechanics yield different answers to the question" (2013: 199).

The discussion about the ontology of EPR is highly relevant in the context of the interpretation of quantum mechanics. However, although entanglement-assisted communication in general, and teleportation in particular, are based on EPR correlations, they are not mere EPR experiments. In fact, in teleportation Alice not only counts with a particle of the entangled pair, but has access also to the state $|\chi\rangle$ to be teleported, and to the two two-state classical systems needed to send the two bits of information through the classical channel. Since communication requires those three elements, the intervention does not need to act on the entangled pair, but it can operate on the other two elements. For instance, the intervention on Alice's end may change the state to be teleported, from $|\chi\rangle$ to $|\varphi\rangle$: as a consequence, something changes in Bob's end, since he will recover $|\varphi\rangle$ and not $|\chi\rangle$. Or the intervention might block one of the classical systems that Alice sends to Bob: in this case, Bob would be unable to recover the teleported state. It is worth stressing that we can be sure about the consequences of these interventions independently of whether the entangled pair is interpreted as a single holistic system.

Although interventionism gives a clear answer to the charge of anthropocentrism directed to the first versions of manipulabilism, in the application of this approach to interpreting the concept of information anthropocentrism is not an issue: here we are not interested in the moon causing tides or in the motion of tectonic plates causing earthquakes. Our context of interest is very limited, since confined to cases of communication, in which there is always a deliberate intervention on the source of information with the purpose to change the state of the destination. Moreover, Cartwright's worries are beyond our limited scope: the fact that the interventionist concept of cause cannot be applied in certain relevant causal situations is not a problem if those situations do not involve communication. In our case, causation is used only as a probe tool to know whether there is transmission of information in the communicational context, independently of signals and interactions between source and destination.

Regarding the objection to the use of counterfactuals, it is also innocuous in our context of interest. In fact, counterfactuals are introduced in the interventionist approach to deal with cases where the intervention on the cause is physically or practically impossible. But in situations of transmission of information, the interventions on the source are always physically and practically possible. Even more, since the messages to be transmitted are embodied in sequences of the states of the source, the possibility of controlling the state of the source is an essential requirement for communication: the nature of communication itself includes that possibility; it makes no sense to conceive a source of information whose state cannot be modified.

It is worth emphasizing again that here it is not argued that the manipulability approach is the right or the best theory of causation, which accounts for all conceivable cases of causation. The discussion of this point is not relevant for our argumentation, since our aim is more limited: we appeal to a manipulability view of causation only to disentangle the problems related to the interpretation of the concept of information in the communicational context.

Summing up, the interventionist version of the manipulability approach to causation seems to be a productive resource to cope with the problem of the interpretation of information in the communicational context in the face of the difficulties introduced by quantum-assisted communication. It agrees with the intuitive idea that communication not only involves correlations, but also essentially requires the possibility of changing the state of the destination by manipulating the state of the source. But, at the same time, it eludes the bizarre answers given by those who try to retain for teleportation an explanation in terms of physical interactions or signals traveling through space and time.

7 Conclusions

Among the many different discursive domains in which the concept of information plays a significant role, this work has focused on the communicational context, perhaps the traditional context to talk about information, where two interpretations of the concept were distinguished. It has been shown that the epistemic interpretation of information, although possibly useful in scientific or philosophical studies about human knowledge, does not incorporate the essential asymmetric feature of communication: what happens in the source of information must produce modifications on what happens in the destination. The physical interpretation does not suffer from this shortcoming to the extent that it conceives information as something physical, that can be generated at one point of the physical space and transmitted to another point, that can be accumulated, stored, and converted from one form to another. From this traditional viewpoint, information is always transferred by means of a physical substratum: source and destination must be linked through a physical signal that carries the information. However, this interpretation faces serious difficulties to deal with teleportation, the typical case studied in the field of quantum-assisted communication: although Alice succeeds in the transmission of the information corresponding to the teleported state, there is no carrier of that information that travels from the source to the destination following a continuous space-time path. For this reason, some authors decided to reject the physical interpretation and to adopt a deflationary view of information, which dissolves the problem of the interpretation of the concept.

The challenge of this work has been to supply a view of information that can overcome these difficulties. It has been argued that a way to reach this purpose is to exploit a basic and intuitive idea behind the concept of communication: the idea of causation, the intuition that a change in the state of the source must produce a change in the destination. Of course, no concept of causation based on physical interactions is useful to that purpose, since it would amount to come back to the traditional physical interpretation of information. By contrast, the manipulability approach to causation, in its interventionist version, seems to be the appropriate conceptual resource to deal with the problem. From this perspective, the essential feature of causation is the capability of controlling the effect by manipulating the cause, independently of whether there is a physical interaction between the cause and the effect. An interpretation of information tied to this view of causation seems to be the suitable tool for picking up all the cases usually conceived as communication situations and only them.

In short, if it is accepted that "no information without causation," and causation is conceived in manipulabilist terms, the slogan becomes "*no information without manipulation.*"

Acknowledgments

We are extremely grateful to all the participants in the workshop "What Is Quantum Information?" (Buenos Aires, May 2015), Jeffrey Bub, Adán Cabello, Dennis Dieks, Armond Duwell, Christopher Fuchs, Robert Spekkens, and Christopher Timpson, for the stimulating and lively discussions about the concept of information. This chapter was partially supported by a Large Grant of the Foundational Questions Institute (FQXi), and by grants of the National Council of Scientific and Technological Research (CONICET) and the National Agency for Scientific and Technological Promotion (ANPCyT-FONCYT) of Argentina.

References

Adriaans, P. (2013). "Information." In E. N. Zalta (ed.), *The Stanford Encyclopedia of Philosophy* (Fall 2013 Edition), http://plato.stanford.edu/archives/fall2013/entries/information/.

Bar-Hillel, Y. (1964). *Language and Information: Selected Essays on Their Theory and Application*. Reading, MA: Addison-Wesley.

Bar-Hillel, Y. and Carnap, R. (1953). "Semantic Information." *The British Journal for the Philosophy of Science*, **4**: 147–157.

Brukner, Č. and Zeilinger, A. (2009). "Information Invariance and Quantum Probabilities." *Foundations of Physics*, **39**: 677–689.

Cartwright, N. (1979). "Causal Laws and Effective Strategies." *Noüs*, **13**: 419–437.

Cartwright, N. (2002). "Against Modularity, the Causal Markov Condition, and Any Link Between the Two: Comments on Hausman and Woodward." *British Journal for the Philosophy of Science*, **53**: 411–453.

Castañeda, H.-N. (1984). "Causes, Causity, and Energy." Pp. 17–27 in P. French, T. Uehling Jr., and H. Wettstein (eds.), *Midwest Studies in Philosophy IX*. Minneapolis: University of Minnesota Press.

Chaitin, G. (1966). "On the Length of Programs for Computing Binary Sequences." *Journal of the Association for Computing Machinery*, **13**: 547–569.

Cook, T. and Campbell, D. (1979). *Quasi-experimentation: Design and Analysis Issues for Field Settings*. Boston, MA: Houghton Mifflin Company.

Cover, T. and Thomas, J. (1991). *Elements of Information Theory*. New York: John Wiley & Sons.

Deutsch, D. and Hayden, P. (2000). "Information Flow in Entangled Quantum Systems." *Proceedings of the Royal Society of London A*, **456**: 1759–1774.

Dowe, P. (1992). "Wesley Salmon's Process Theory of Causality and the Conserved Quantity Theory." *Philosophy of Science*, **59**: 195–216.

Dretske, F. (1981). *Knowledge and the Flow of Information*. Oxford: Basil Blackwell.

Dunn, J. (2001). "The Concept of Information and the Development of Modern Logic." Pp. 423–427 in W. Stelzner (ed.), *Non-classical Approaches in the Transition from Traditional to Modern Logic*. Berlin: de Gruyter.

Duwell, A. (2008). "Quantum Information Does Exist." *Studies in History and Philosophy of Modern Physics*, **39**: 195–216.

Ehring, D. (1986). "The Transference Theory of Causality." *Synthese*, **67**: 249–258.

Einstein, A., Podolsky, B., and Rosen, N. (1935). "Can Quantum-Mechanical Description of Physical Reality be Considered Complete?" *Physical Review*, **47**: 777–780.

Fair, D. (1979). "Causation and the Flow of Energy." *Erkenntnis*, **14**: 219–250.

Fisher, R. (1925). "Theory of Statistical Estimation." *Proceedings of the Cambridge Philosophical Society*, **22**: 700–725.

Floridi, L. (2011). *The Philosophy of Information*. Oxford: Oxford University Press.

Frisch, M. (2014). *Causal Reasoning in Physics*. Cambridge: Cambridge University Press.

Hausman, D. and Woodward, J. (1999). "Independence, Invariance, and the Causal Markov Condition." *British Journal for the Philosophy of Science*, **50**: 521–583.

Hausman, D. and Woodward, J. (2004). "Modularity and the Causal Markov Condition: A Restatement." *The British Journal for the Philosophy of Science*, **55**: 147–161.

Hiddleston, E. (2005). "Review of *Making Things Happen*. " *Philosophical Review*, **114**: 545–547.

Hoel, E., Albantakis, L., and Tononi, G. (2013). "Quantifying Causal Emergence Shows that Macro Can Beat Micro." *Proceedings of the National Academy of Sciences*, **110**: 19790–19795.

Holevo, A. (1973). "Information Theoretical Aspects of Quantum Measurement." *Problems of Information Transmission (USSR)*, **9**: 177–183.

Jozsa, R. (1998). "Quantum Information and Its Properties." Pp. 49–75 in H.-K. Lo, S. Popescu, and T. Spiller (eds.), *Introduction to Quantum Computation and Information*. Singapore: World Scientific.

Khinchin, A. (1957). *Mathematical Foundations of Information Theory*. New York: Dover.

Kistler, M. (1998). "Reducing Causality to Transmission." *Erkenntnis*, **48**: 1–24.

Kolmogorov, A. (1965). "Three Approaches to the Quantitative Definition of Information." *Problems of Information Transmission*, **1**: 4–7.

Kolmogorov, A. (1968). "Logical Basis for Information Theory and Probability Theory." *Transactions on Information Theory*, **14**: 662–664.

Landauer, R. (1991). "Information Is Physical." *Physics Today*, **44**: 23–29.

Landauer, R. (1996). "The Physical Nature of Information." *Physics Letters A*, **217**: 188–193.

Lombardi, O. (2004). "What Is Information?" *Foundations of Science*, **9**: 105–134.

Lombardi, O. (2005). "Dretske, Shannon's Theory and the Interpretation of Information." *Synthese*, **144**: 23–39.

Lombardi, O., Fortin, S., and López, C. (2016). "Deflating the Deflationary View of Information." *European Journal for Philosophy of Science*, **6**: 209–230.

Lombardi, O., Fortin, F., and Vanni, L. (2015). "A Pluralist View about Information." *Philosophy of Science*, **82**: 1248–1259.

Lombardi, O., Holik, F., and Vanni, L. (2016a). "What Is Quantum Information?" *Studies in History and Philosophy of Modern Physics*, **56**: 17–26.

Lombardi, O., Holik, F., and Vanni, L. (2016b). "What Is Shannon Information?" *Synthese*, **193**: 1983–2012.

MacKay, D. (1956). "Towards an Information-Flow Model of Human Behaviour." *British Journal of Psychology*, **47**: 30–43.

MacKay, D. (1969). *Information, Mechanism and Meaning*. Cambridge, MA: MIT Press.

Menzies, P. and Price, H. (1993). "Causation as a Secondary Quality." *British Journal for the Philosophy of Science*, **44**: 187–203.

Nauta, D. (1972). *The Meaning of Information*. The Hague: Mouton.

Nielsen, M. and Chuang, I. (2010). *Quantum Computation and Quantum Information*. Cambridge: Cambridge University Press.

Pearl, J. (2000). *Causality: Models, Reasoning and Inference*. New York: Cambridge University Press.

Penrose, R. (1998). "Quantum Computation, Entanglement and State Reduction." *Philosophical Transactions of the Royal Society of London A*, **356**: 1927–1939.

Price, H. (1991). "Agency and Probabilistic Causality." *British Journal for the Philosophy of Science*, **42**: 157–176.

Reza, F. (1961). *Introduction to Information Theory*. New York: McGraw-Hill.

Rovelli, C. (1996). "Relational Quantum Mechanics." *International Journal of Theoretical Physics*, **35**: 1637–1678.

Russell, B. (1912). "On the Notion of Cause." *Proceedings of the Aristotelian Society*, **13**: 1–26.

Russell, B. (1948). *Human Knowledge: Its Scope and Limits*. New York: Simon and Schuster.

Schumacher, B. (1995). "Quantum Coding." *Physical Review A*, **51**: 2738–2747.

Shannon, C. (1948). "The Mathematical Theory of Communication." *Bell System Technical Journal*, **27**: 379–423, 623–656.

Shannon, C. and Weaver, W. (1949). *The Mathematical Theory of Communication*. Urbana and Chicago: University of Illinois Press.

Solomonoff, R. (1964). "A Formal Theory of Inductive Inference." *Information and Control*, **7**: 1–22, 224–254.

Stonier, T. (1990). *Information and the Internal Structure of the Universe: An Exploration into Information Physics*. New York-London: Springer.

Stonier, T. (1996). "Information as a Basic Property of the Universe." *Biosystems*, **38**: 135–140.

Suárez, M. (2013). "Interventions and Causality in Quantum Mechanics." *Erkenntnis*, **78**: 199–213.

Timpson, C. (2004). *Quantum Information Theory and the Foundations of Quantum Mechanics*. PhD diss., University of Oxford (quant-ph/0412063).

Timpson, C. (2006). "The Grammar of Teleportation." *The British Journal for the Philosophy of Science*, **57**: 587–621.

Timpson, C. (2013). *Quantum Information Theory and the Foundations of Quantum Mechanics*. Oxford: Oxford University Press.

Von Wright, G. (1971). *Explanation and Understanding*. Ithaca, NY: Cornell University Press.

Woodward, J. (2003). *Making Things Happen: A Theory of Causal Explanation*. Oxford: Oxford University Press.

Woodward, J. (2007). "Causation with a Human Face." Pp. 66–105 in H. Price and R. Corry (eds.), *Causation, Physics, and the Constitution of Reality: Russell's Republic Revisited*. Oxford: Oxford University Press.

Woodward, J. (2013). "Causation and Manipulability." E. N. Zalta (ed.), *The Stanford Encyclopedia of Philosophy* (Winter 2013 Edition), http://plato.stanford.edu/archives/win2013/entries/causation-mani/.

Zeilinger, A. (1999). "A Foundational Principle for Quantum Mechanics." *Foundations of Physics*, **29**: 631–643.

Part II

Information and Quantum Mechanics

4

Quantum versus Classical Information

JEFFREY BUB

1 Introduction

The question 'What is quantum information?' has two parts. Firstly, what is information? Secondly, what is the difference between quantum information and classical information? Here I propose an answer to the second question: quantum information is a type of information that is only possible in a world in which there are intrinsically random events.

The quantum world is characterized by the existence of nonlocal correlations that violate Bell inequalities. Such correlations, which do not occur in a classical world, require that the correlated events are separately intrinsically random or 'free' in the sense that they are uncorrelated with any prior events. The nonclassical information-theoretic structure of the quantum world – the possibilities for representing, manipulating, and communicating information – is a manifestation of this intrinsic randomness.

2 Intrinsic Randomness

What is an 'intrinsically random' event? It should be an event that is independent of any information available before the event occurs. In a relativistic universe, different inertial observers will disagree about time order. I'll follow a proposal by Colbeck and Renner (2012) that an event is intrinsically random relative to a time order, and relative to a specific event in this time order.

In special relativity, the time order is defined by the lightcone structure. For a pair of entangled photons, the outcomes of polarization measurements on the photons can be shown to be intrinsically random with respect to the preparation of the entangled state, in the sense that they are independent of any events at all that occur, in any reference frame, before the preparation of the entangled state. Putting it differently, they are uncorrelated with any event that is not in the future lightcone of the preparation event, so independent of any event that the preparation itself couldn't have caused. The outcome of a polarization measurement on one photon is predictable from the outcome of a polarization measurement on the second photon, and conversely, but the polarization measurements and their outcomes are events that occur after the preparation of the entangled state in every reference frame.

For two photons in a maximally entangled state, Colbeck and Renner (2011) use a chained Bell inequality to show that there can't be a variable, associated with the history of the photons before the preparation of the entangled state in the reference frame of any inertial observer, that provides information about the outcomes of polarization measurements on the photons, so that Alice's and Bob's marginal probabilities conditional on this variable are closer to 1 than the probabilities of the entangled state. In other words, the information in the entangled state about the probabilities of measurement outcomes is as complete as it could be. Next, they show that the same conclusion follows for any entangled state, not necessarily a maximally entangled state. Finally, they extend the argument to any quantum state by exploiting the fact that a measurement interaction described by the dynamics of quantum mechanics leads to an entangled state of the measured system and the measuring instrument.

Here's a simple way to see that measurement outcomes on entangled photons are intrinsically random. As a preliminary result, I'll first show that the outputs of a hypothetical Popescu-Rohrlich (PR) box for specific inputs are intrinsically random. A PR box has two parts, which I'll call the Alice part and the Bob part. Each part has two inputs, which can each be 0 or 1, and two outputs for each input, also 0 or 1. The Alice part and the Bob part can be pulled apart by any distance without altering the correlation, which I'll take as: the outputs are different for the input pair 01, and the same for the input pairs 00, 10, 11, with the first element in each pair representing Alice's input and the second element representing Bob's input. This particular correlation can be expressed by $a \oplus b = (A \oplus 1) \cdot B$, where A, a represent Alice's input and output, respectively, and B, b represent Bob's input and output. (There are eight possible PR correlations, four depending on which pair of inputs is associated with different outputs, and four more where the outputs are the same for one pair of inputs, and different for three pairs of inputs.)

I'll use the no-signalling principle (Alice's marginal probabilities are independent of Bob's actions, or whether Bob does anything at all, and conversely) and a chain of equalities to show that the marginal probabilities for Alice and Bob separately are all 1/2. This follows from the PR correlation even if the probabilities are conditional on events that occur, in any reference frame, before the PR box comes into existence. So the output of a PR box for a particular input is intrinsically random with respect to the creation of the PR box, in the sense that the outputs are independent of any events at all that occur, in any reference frame, before the correlation of the PR box came into existence.

From no signalling, the probability that Alice's output is 0 for her input 0 doesn't depend on Bob's input. So this probability is just the sum of the probabilities that Alice's output is 0 for the two possible outputs for Bob's input 0:

$$p(0_A|0_A) = p(0_A 0_B|00) + p(0_A 1_B|00)$$

(For clarity here, Alice's and Bob's outputs are indicated by the appropriate subscripts.) Similarly, the probability that Bob's output is 0 for his input 0 is the sum of the probabilities for the two possible outputs for Alice's input 0:

$$p(0_B|0_B) = p(0_A 0_B|00) + p(1_A 0_B|00)$$

The first terms in the sums on the right-hand sides of these two equations are the same and the second terms are zero (because the outputs are the same for the pair of inputs 00), so

$$p(0_A|0_A) - p(0_B|0_B) = 0$$

Similarly,

$$p(0_B|0_B) - p(0_A|1_A) = 0$$

$$p(0_A|1_A) - p(0_B|1_B) = 0$$

$$p(0_B|1_B) - p(1_A|0_A) = 0$$

These four equations are 'chained' in the sense that the second term on the left-hand side of each equation, except the last, is the same as the first term on the left-hand side of the following equation, but with a minus sign. The sum of the terms on the left-hand sides is equal to the sum of the terms on the right-hand sides, which is zero. So from this chain of four equations you get

$$p(0_A|0_A) - p(1_A|0_A) = 0$$

Since $p(1_A|0_A) = 1 - p(0_A|0_A)$, it follows that $p(0_A|0_A) = 1/2$ and so $p(1_A|0_A) = 1/2$. The same reasoning shows that $p(0_A|1_A) = p(1_A|1_A) = 1/2$, and similarly for Bob's marginal probabilities. The correlation array in Table 4.1 shows the probabilities for this PR box.

Suppose e represents any event that occurs, in any reference frame, before the PR box comes into existence. If you conditionalize the probabilities in the previous

Table 4.1 *The Popescu-Rohrlich correlation array for the correlation* $a \oplus b = (A \oplus 1) \cdot B$.

Alice		0		1	
Bob		0	1	0	1
0	0	1/2	0	1/2	0
	1	0	1/2	0	1/2
1	0	0	1/2	1/2	0
	1	1/2	0	0	1/2

Table 4.2 *The quantum correlation array for pairs of photons in the state* $|\phi^+\rangle$ *when Alice measures polarization in direction* $A = 0$, $A' = \pi/4$, *and Bob measures polarization in directions* $B = \pi/8$, $B' = 3\pi/8$.

Alice		A		A'	
Bob		0	1	0	1
B	0	$\frac{1}{2}cos^2\frac{\pi}{8}$	$\frac{1}{2}sin^2\frac{\pi}{8}$	$\frac{1}{2}cos^2\frac{\pi}{8}$	$\frac{1}{2}sin^2\frac{\pi}{8}$
	1	$\frac{1}{2}sin^2\frac{\pi}{8}$	$\frac{1}{2}cos^2\frac{\pi}{8}$	$\frac{1}{2}sin^2\frac{\pi}{8}$	$\frac{1}{2}cos^2\frac{\pi}{8}$
B'	0	$\frac{1}{2}sin^2\frac{\pi}{8}$	$\frac{1}{2}cos^2\frac{\pi}{8}$	$\frac{1}{2}cos^2\frac{\pi}{8}$	$\frac{1}{2}sin^2\frac{\pi}{8}$
	1	$\frac{1}{2}cos^2\frac{\pi}{8}$	$\frac{1}{2}sin^2\frac{\pi}{8}$	$\frac{1}{2}sin^2\frac{\pi}{8}$	$\frac{1}{2}cos^2\frac{\pi}{8}$

argument on e, the equations all remain true, if Alice and Bob can't violate the no-signalling principle by exploiting information about e. For example, $p(0_A|0_A, e)$ doesn't depend on Bob's input, so

$$p(0_A|0_A, e) = p(0_A 0_B|00, e) + p(0_A 1_B|00, e)$$

and so on. It follows that the marginal probabilities are independent of e because they remain equal to 1/2 and are not altered by conditionalizing on e.

Now suppose Alice and Bob share many copies of entangled photons in the maximally entangled state $|\phi^+\rangle = 1/\sqrt{2}(|0\rangle|0\rangle + |1\rangle|1\rangle)$. Alice measures the polarization of her photons in directions $A = 0$ or $A' = \pi/4$, and Bob measures the polarization of his photons in directions $B = \pi/8$ or $B' = 3\pi/8$. Table 4.2 shows the resulting correlation array, with 0 and 1 representing measurement outcomes 'horizontal' and 'vertical'. To compare with the Popescu-Rohrlich probabilities of 0 and 1/2 in Table 4.1, the approximate values of the probabilities for $|\phi^+\rangle$ are

Table 4.3 *Approximate values for the probabilities in Table 4.2.*

	Alice	A		A'	
Bob		0	1	0	1
B	0	0.427	0.073	0.427	0.073
	1	0.073	0.427	0.073	0.427
B'	0	0.073	0.427	0.427	0.073
	1	0.427	0.073	0.073	0.427

Table 4.4 *Deterministic correlation arrays D_1 and D_2.*

	Alice	A		A'	
Bob		0	1	0	1
B	0	1	0	1	0
	1	0	0	0	0
B'	0	1	0	1	0
	1	0	0	0	0

	Alice	A		A'	
Bob		0	1	0	1
B	0	0	0	0	0
	1	0	1	0	1
B'	0	0	0	0	0
	1	0	1	0	1

shown in Table 4.3. The entries $(1/2)\cos^2(\pi/8)$ and $(1/2)\sin^2(\pi/8)$ in the correlation array in Table 4.2 are approximately 0.427 and 0.073, respectively, rather than $1/2$ and 0 in the correlation array in Table 4.1. So $p(\text{outcomes different}) = \sin^2(3\pi/8) = \cos^2(\pi/8) \approx .854$ for the case AB' when the angle between the polarizations is $3\pi/8$, $p(\text{outcomes same}) = \cos^2(\pi/8) \approx .854$ for the three cases AB, $A'B$, $A'B'$ when the angle between the polarizations is $\pi/8$. These correspond to the PR probabilities $p(\text{outputs different}) = 1$ for input 11 and $p(\text{outputs same}) = 1$ for the inputs 00, 10, 11. The photon correlation is as close as you can get with entangled photons to the PR correlation in Table 4.1.

The correlation array in Table 4.2 can be expressed as a mixture or probability distribution of the deterministic correlation arrays D_1 and D_2 in Table 4.4, and D_3 and D_4 in 4.5, each weighted with a probability $(1/2)\sin^2(\pi/8)$, and the PR correlation array in Table 4.1, weighted with probability $1 - 2\sin^2(\pi/8)$. Symbolically,

$$\frac{1}{2}\sin^2\frac{\pi}{8}(D_1 + D_2 + D_3 + D_4) + \left(1 - 2\sin^2\frac{\pi}{8}\right)PR$$

where PR here stands for the correlation array in Table 4.1.

Table 4.5 *Deterministic correlation arrays* D_3 *and* D_4.

<table>
<tr><td rowspan="2">Alice
Bob</td><td colspan="2">A</td><td colspan="2">A'</td></tr>
<tr><td>0</td><td>1</td><td>0</td><td>1</td></tr>
<tr><td>B 0</td><td>0</td><td>0</td><td>0</td><td>0</td></tr>
<tr><td>   1</td><td>1</td><td>0</td><td>1</td><td>0</td></tr>
<tr><td>B' 0</td><td>0</td><td>0</td><td>0</td><td>0</td></tr>
<tr><td>   1</td><td>1</td><td>0</td><td>1</td><td>0</td></tr>
</table>

<table>
<tr><td rowspan="2">Alice
Bob</td><td colspan="2">A</td><td colspan="2">A'</td></tr>
<tr><td>0</td><td>1</td><td>0</td><td>1</td></tr>
<tr><td>B 0</td><td>0</td><td>1</td><td>0</td><td>1</td></tr>
<tr><td>   1</td><td>0</td><td>0</td><td>0</td><td>0</td></tr>
<tr><td>B' 0</td><td>0</td><td>1</td><td>0</td><td>1</td></tr>
<tr><td>   1</td><td>0</td><td>0</td><td>0</td><td>0</td></tr>
</table>

To see this, for each of the 16 entries in Table 4.2, add the values of the entries in the corresponding positions in the four deterministic correlation arrays and the PR correlation array in Table 4.1, weighted in the appropriate way. So, for example, adding the appropriately weighted values for the outcome 00 for the measurement pair AB (the entry in the top left position in the top left cell) gives $(1/2) \sin^2(\pi/8) \times 1 + (1/2) \sin^2(\pi/8) \times 0 + (1/2) \sin^2(\pi/8) \times 0 + (1/2) \sin^2(\pi/8) \times 0 + \left(1 - 2 \sin^2(\pi/8)\right) \times 1/2 = (1/2)\left(1 - \sin^2(\pi/8)\right) = (1/2) \cos^2(\pi/8)$ and similarly for the other entries.

The deterministic correlations defined by the arrays D_1, D_2, D_3, D_4 are local correlations. Alice and Bob could simulate the correlation D_1 by both responding 0 for A, A', B, B', and they could simulate the correlation D_2 by both responding 1 for A, A', B, B'. To simulate the correlation D_3, Alice responds 0 for A, A' and Bob responds 1 for B, B', and to simulate the correlation D_4 Alice responds 1 for A, A' and Bob responds 0 for B, B'. If Alice and Bob had access to the PR box represented by Table 4.1, they could simulate the correlations of the entangled quantum state $|\phi^+\rangle$ with a combination of local resources corresponding to the deterministic correlations D_1, D_2, D_3, D_4 and a nonlocal intrinsically random resource corresponding to the PR box. They could do this by sharing a list of random numbers $1, \ldots, 5$, where the numbers $1, \ldots, 4$ occur with equal probability $(1/2) \sin^2(\pi/8)$ and 5 occurs with probability $1 - 2 \sin^2(\pi/8)$. For the numbers $1, \ldots, 4$ they respond according to the deterministic array D_1, D_2, D_3, or D_4, and for the number 5 they respond according to PR box array in Table 4.1, taking A, A' and B, B' as the PR box inputs 0, 1.

It follows from Bell's locality argument (1964) that the correlation in Table 4.2 can't be simulated with local resources. In a world with PR boxes, the correlation can be simulated with a mixture of local resources and $1 - 2 \sin^2(\pi/8) \approx 70\%$ intrinsically random PR-boxes, a nonlocal resource. Since the outputs of a PR box for specific inputs are intrinsically random, you could say that, in a world with PR boxes, about 70 per cent of the measurement outcomes for the quantum correlation are intrinsically random. In our quantum world there are no PR correlations to mix

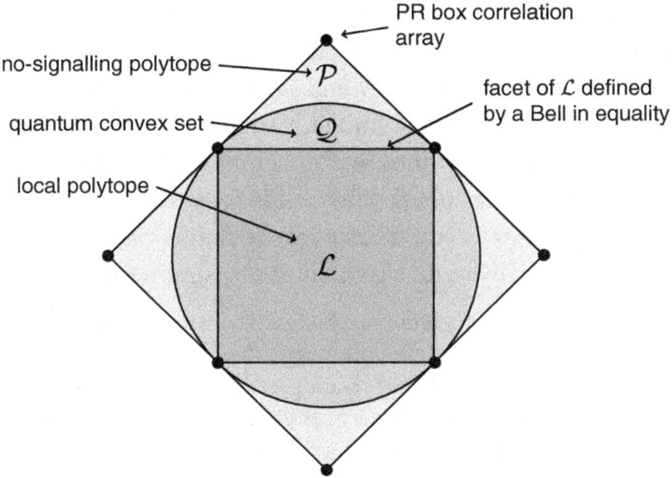

Figure 4.1. A schematic representation of local and nonlocal correlations.

with deterministic correlations to get the photon correlation, so measurement outcomes on the photons are intrinsically random.

The relation between classical, quantum, and superquantum correlations can be represented schematically by three nested sets (Figure 4.1). For the case considered earlier, each point in the space represents a correlation array of 16 probabilities, with two inputs or measurements and two outputs or measurement outcomes for each input or measurement for Alice and for Bob. There are four independent probability normalization constraints (the probabilities in each quadrant of the correlation array must add to 1) and four independent no-signalling constraints. This reduces the number of independent probability variables to eight, so the space is eight-dimensional. The local polytope $\mathcal{L}$, whose vertices or extremal points represent deterministic correlation arrays, includes all and only correlations which have a common cause explanation. These correlations can be generated by classical probability distributions over the values of variables representing the common causes of the correlations, which are not explicit in the description of the phenomenal correlations represented by points of the local polytope and so are 'hidden variables'. The Bell inequalities define the faces of the local polytope. Points outside the local polytope represent nonlocal correlations: quantum correlations in the convex set $\mathcal{Q}$ (whose boundary is not hyperspherical in the eight-dimensional space as the schematic diagram suggests, but a complicated continuous manifold limited by the Tsirelson bound) and superquantum correlations between $\mathcal{Q}$ and the outermost no-signalling polytope $\mathcal{P}$. The vertices of $\mathcal{P}$ include the eight PR boxes, which are extremal no-signalling correlation arrays.

The nonlocal probabilistic correlations in the region between the boundary of the local polytope $\mathcal{L}$ and the boundary of the no-signalling polytope $\mathcal{P}$ are, as von Neumann (1962) put it, 'perfectly new and *sui generis* aspects of physical reality' and don't represent ignorance about hidden variables that have been left out of the story. One could say that correlations in $\mathcal{L}$ are completely explained by the systems having what Einstein called a 'being-thus', an independent existence characterized by definite properties, prior to any measurement, that is associated, either deterministically or probabilistically, with a particular measurement outcome (1948: 170):

If one asks what, irrespective of quantum mechanics, is characteristic of the world of ideas of physics, one is first of all struck by the following: the concepts of physics relate to a real outside world, that is, ideas are established relating to things such as bodies, fields, etc., which claim a 'real existence' that is independent of the perceiving subject – ideas which, on the other hand, have been brought into as secure a relationship as possible with sense-data. It is further characteristic of these physical objects that they are thought of as arranged in a physical space-time continuum. An essential aspect of this arrangement of things in physics is that they lay claim, at a certain time, to an existence independent of one another, provided that these objects 'are situated in different parts of space'. Unless one makes this kind of assumption about the independence of the existence (the 'being-thus') of objects which are far apart from one another in space – which stems in the first place from everyday thinking – physical thinking in the familiar sense would not be possible. It is also hard to see any way of formulating and testing the laws of physics unless one makes a clear distinction of this kind.

Correlations outside $\mathcal{L}$ can't be simulated with local resources, so such correlated systems can't have a 'being-thus' in the sense Einstein had in mind. To be sure, 'physical thinking in the familiar sense' would have to change to accommodate the existence of nonlocal correlations outside $\mathcal{L}$, but formulating and testing a physical theory like quantum mechanics that produces correlations outside $\mathcal{L}$ doesn't present any special problem. The phenomena are objective. We don't know how the universe makes composite systems with nonlocal correlations, but we don't know how the universe makes objects with independent 'being-thuses' either. We simply take elementary objects with elementary 'being-thuses' as not requiring further explanation in classical physics. What classical physics tells us is how these elementary objects combine to form complex objects, and how these objects transform dynamically. Given elementary objects with elementary 'being-thuses' as primitive entities, we can understand how classical correlations arise from common or direct causes. The situation outside $\mathcal{L}$ is not really different in this respect. We can understand how nonlocal correlations arise in our quantum world if we take indefinite systems that become definite in an intrinsically random way when measured as elementary entities not requiring further explanation. What does change radically are the possibilities for representing, manipulating, and communicating information by exploiting correlations outside $\mathcal{L}$ – the structure of information inside and outside $\mathcal{L}$ is very different.

3 Quantum Information

In the previous section I showed that the nonlocal correlations of entangled quantum systems are only possible if the measurement outcomes on the separate systems are intrinsically random events. In a review article in *Nature Physics*, Sandu Popescu remarks (2014: 264):

> Indeed, very roughly speaking, if by moving something here, something else instantaneously wiggles there, the only way in which this doesn't lead to instantaneous communication is if that 'wiggling thing' is uncertain and the wiggling can only be spotted *a posteriori*.

The 'wiggling can only be spotted *a posteriori*' because nothing changes for Bob when Alice does something to her photon, but if Alice communicates to Bob what she did (which requires a telephone call or other form of subluminal communication), Bob can confirm this by measuring the polarization of his photon in an appropriate direction.

One might think that intrinsic randomness would reduce the possibilities for information-processing, but the opposite is the case, essentially because new sorts of correlations are possible that cannot occur in a classic world. If Alice and Bob had access to PR boxes, they could enhance the capacity of classical computers to perform information-theoretic tasks, beyond the information-theoretic possibilities inherent in classical computers. As an example, Popescu cites the 'calendar problem', an instance of the inner product problem, which is the most difficult communication task possible. Alice and Bob, who are in different cities, want to find out whether the number of days they could meet in a year is even or odd (taking 0 as even). How much information must Bob send to Alice for her to be able to answer the question? She can answer the question if Bob sends her 365 bits of information about whether he is free, one bit for each day of the year. With this solution to the problem, Alice ends up with a lot of redundant information, since all she is interested in is the one bit of information that is the answer to the question: even or odd. Popescu shows how Alice and Bob can solve the problem with a nonlocal resource: a supply of 365 PR boxes. If each PR box (represented by $a \oplus b = A \cdot B$ in this case) is associated with a day of the year, Alice and Bob each input 0 or 1 into their part of the box, depending on whether they are free on that day. For each PR box, the Boolean sum of the outputs is equal to the product of the inputs, so the sum of the outputs $\sum_i a_i \oplus b_i$ for all 365 PR boxes is even if and only if the sum of the input products $\sum_i A_i \cdot B_i$ is even. It follows that all Alice needs to know to answer the question is whether the sum of the outputs to Bob's PR boxes, $\sum_i b_i$, is even or odd, which is just one bit of information.

Since any communication problem can be mapped onto the inner product problem, the existence of PR boxes would trivialize communication complexity:

all communication problems could be solved with one bit of communication. This seems too good to be true, so one might suspect that some information-theoretic principle characterizing our quantum world excludes PR boxes, and the search for such a principle is an active area of research. Entangled quantum states are, roughly speaking, about 85 per cent like PR boxes: these correlations are closer to PR box correlations than classical correlations, which Bell's locality argument shows can achieve a success rate of no more than 75 per cent in any simulation. It turns out that for this particular information-theoretic task, entanglement doesn't help, but there are information-theoretic tasks for which it does.

The transition from classical to relativistic physics rests on Einstein's insight that there is a structure to space-time and this structure is not what Newton thought it was. What distinguishes the quantum world from the classical world is the existence of nonlocal correlations that can only occur between intrinsically random events, and this means that the information-theoretic structure of the quantum world is not what Shannon thought it was. As Schrödinger remarked (1935: 555), entanglement is '*the characteristic trait of quantum mechanics, the one that enforces its entire departure from classical lines of thought*'. The entanglement of composite systems is an immediate consequence of Heisenberg's *Umdeutung* (1925), his 'reinterpretation' of kinematical and mechanical relations as noncommutative, the core step from classical to quantum mechanics. More precisely, it is a consequence of the way in which commuting and noncommuting observables are 'intertwined' in quantum mechanics, to use Gleason's term (1957: 886).

Shannon was interested in how to quantify the minimal physical resources required to store messages produced by a source, so that they could be communicated via a channel without loss and reconstructed by a receiver. Shannon's source coding theorem or noiseless channel coding theorem (1948) answers this question.

The idea is to consider a source that produces long sequences (messages) composed of symbols from a finite alphabet $a_1, a_2, \ldots, a_k$, where the individual symbols are produced with probabilities $p_1, p_2, \ldots, p_k$. A given sequence of symbols is represented as a sequence of values of independent, identically distributed, discrete, random variables $X_1, X_2, \ldots$. A typical sequence of length n, for large n, will contain close to $p_i n$ symbols a_i, for $i = 1, \ldots, n$. The probability of a sufficiently long typical sequence is

$$p(x_1, x_2, \ldots, x_n) = p(x_1)p(x_2) \ldots p(x_n) \approx p_1^{p_1 n} p_2^{p_2 n} \ldots p_k^{p_k n}$$

so

$$\log p(x_1, x_2, \ldots, x_n) \approx n \sum_i p_i \log p_i := -nH(X)$$

where $H(X) := -\sum_i p_i \log p_i$ is the Shannon entropy of the source (where the convention in information theory is to take the logarithm to the base 2). If we take $-\log p_i$, a decreasing function of p_i with a minimum value of 0 when $p_i = 1$ for some i, as a measure of the information associated with identifying the symbol a_i produced by an information source, then $H(X) := -\sum_i p_i \log p_i$ is the average information gain associated with ascertaining the value of the random variable X.

Since

$$p(x_1, x_2, \ldots, x_n) = 2^{-nH(X)}$$

for sufficiently long typical sequences, and the probability of all the typical n-length sequences is less than 1, it follows that there are at most $2^{-nH(X)}$ typical sequences. Since the probability that the source produces an atypical sequence tends to zero as $n \to \infty$, the set of typical sequences has probability close to 1. So each typical n-sequence can be encoded as a distinct binary number of $nH(X)$ binary digits or bits before being sent through the channel to the receiver, where the original sequence can then be reconstructed by inverting the 1–1 encoding map. The reconstruction would fail, with low probability, only for the rare atypical sequences, each of which could be encoded as, say, a string of 0's.

If the probabilities p_i are all equal ($p_i = 1/k$ for all i), then $H(X) = \log k$, and if some $p_j = 1$ (and so $p_i = 0$ for $i \neq j$), then $H(X) = 0$ (taking $0 \log 0 = \lim_{x \to 0} x \log x = 0$). So

$$0 \leq H(X) \leq \log k$$

If each of the k distinct symbols is encoded as a distinct binary number, i.e., as a distinct string of 0's and 1's, we would need binary numbers composed of $\log k$ bits to represent each symbol ($2^{\log k} = k$). So Shannon's analysis shows that messages produced by a stochastic source can be compressed, in the sense that, as $n \to \infty$ and the probability of an atypical n-length sequence tends to zero, n-length sequences can be encoded without loss of information using $nH(X)$ bits rather than the $n \log k$ bits required if we encoded each of the k symbols a_i as a distinct string of 0's and 1's. This is a compression, since $nH(X) < n \log k$ except for equiprobable distributions.

The significance of Shannon's source coding theorem lies in showing that there is an optimal or most efficient way of compressing messages produced by a source in such a way that they can be reliably reconstructed by a receiver. Since a message is abstracted as a sequence of distinguishable symbols produced by a stochastic source, the only relevant feature of a message with respect to reliable compression and decompression is the sequence of probabilities associated with the individual symbols: the nature of the physical systems instantiating the representation of the message through their states is irrelevant to this notion of compression (provided only that the states are reliably distinguishable), as is the content or meaning of the message. The Shannon entropy $H(X)$ is a measure of the minimal physical resources, in terms of the average number of bits per symbol, that are necessary and sufficient to reliably store the output of a source of messages. In this sense, it is a measure of the amount of information per symbol produced by an information source.

Schumacher's source coding theorem for quantum information (1995) extends Shannon's notion of compressibility to a stochastic source of quantum states. If a composite quantum system QE is in an entangled state $|\Psi\rangle$, the subsystem Q is in a mixed state ρ, which can be expressed as

$$\rho = \sum_i p_i |i\rangle \langle i|$$

where the p_i are the eigenvalues of the density operator ρ and the pure states $|i\rangle$ are orthonormal eigenstates of ρ. This is the spectral representation of ρ, and any density operator – a positive (hence Hermitian) operator – can be expressed in this way. The representation is unique if and only if the p_i are all distinct. If some of the p_i are equal, there is a unique representation of ρ as a sum of projection operators with the distinct values of the p_i as coefficients, but some of the projection operators will project onto multidimensional subspaces.

Since ρ has unit trace, $\sum p_i = 1$, and so the spectral representation of ρ represents a classical probability distribution of orthogonal, and hence distinguishable, pure states. If we measure a Q-observable with eigenstates $|i\rangle$, the outcomes can be associated with the values of a random variable X, where $\Pr(X = i) = p_i$. Then

$$H(X) = -\sum_i p_i \log p_i$$

is the Shannon entropy of the probability distribution of measurement outcomes. Now

$$-Tr(\rho \log \rho) = -\sum_i p_i \log p_i$$

(because the eigenvalues of $\rho \log \rho$ are $p_i \log p_i$ and the trace of an operator is the sum of the eigenvalues), so a natural generalization of Shannon entropy for any not necessarily orthogonal mixture of quantum states with density operator ρ is the von Neumann entropy

$$S := -Tr(\rho \log \rho)$$

which coincides with the Shannon entropy for measurements in the eigenbasis of ρ. For a completely mixed state $\rho = I/d$, where $\dim \mathcal{H}^Q = d$, the d eigenvalues of ρ are all equal to $1/d$ and $S = \log d$. This is the maximum value of S in a d-dimensional Hilbert space. The von Neumann entropy S is zero, the minimum value, if and only if ρ is a pure state, where the eigenvalues of ρ are 1 and 0. So $0 \leq S \leq \log d$.

The von Neumann entropy does *not*, in general, represent the amount of information gained by identifying the quantum state produced by a stochastic source given by a mixed state, because the quantum states in a mixture prepared by mixing nonorthogonal quantum states cannot be reliably identified. Nonetheless, Schumacher's theorem shows that quantum information, like classical information, can be compressed.

For a classical information source of bits, where the output of the source is given by a random variable X with two possible values x_1, x_2 with probabilities p_1, p_2, the Shannon entropy of the information produced by the source is $H(X) = H(p_1, p_2)$. So by Shannon's source coding theorem the information can be compressed and communicated to a receiver with arbitrarily low probability of error with $H(X)$ bits per signal, which is less than one bit if $p_1 \neq p_2$.

If the source produces qubit states $|\psi_1\rangle$, $|\psi_2\rangle$ with probabilities p_1, p_2, the von Neumann entropy of the mixture $\rho = p_1|\psi_1\rangle\langle\psi_1| + p_2|\psi_2\rangle\langle\psi_2|$ is $S(\rho)$. Schumacher's generalization of Shannon's source coding theorem shows that the quantum information encoded in the mixture ρ can be compressed and communicated to a receiver with arbitrarily low probability of error by using $S(\rho)$ qubits per signal, and $S(\rho) < 1$ if the quantum states are nonorthogonal. So we can reliably communicate the sequence of qubit states produced by the source by sending less than one qubit per signal. Since $S(\rho) < H(p_1, p_2)$ if the qubit states are nonorthogonal, the quantum information represented by the sequence of qubit states can be compressed beyond the maximum compression of the classical information associated with the Shannon entropy of the random variable whose values are the labels of the qubit states.

What does the difference between Schumacher's theorem and Shannon's theorem have to do with intrinsic randomness? In the classical case, the source produces a mixture of distinguishable classical states, each associated with a definite catalogue of properties corresponding to Einstein's 'being-thus'. In the quantum case, the source produces a quantum mixture, which is indefinite with respect to its representation in terms of pure states. Since a 'proper' quantum mixture prepared by mixing certain quantum states is indistinguishable from an 'improper' mixture derived by averaging over ('tracing over') an ancilla system entangled with the source, the information a quantum source produces is the sort of information associated with the intrinsically random outcomes of measurements on entangled systems.

4 Conclusion

The salient difference between our quantum world and the classical world of Newton, or even Einstein, is nonlocality: the existence of nonlocal correlations that are only possible if the correlated events are separately intrinsically random. This intrinsic randomness underlies the possibility of new sorts of information processing that transcend the possibilities inherent in a classical world and marks the difference between quantum and classical information.

Acknowledgements

The research for this chapter was supported in part by the University of Maryland Institute for Physical Science and Technology and the Joint Center for Quantum Information and Computer Science.

References

Bell, J. S. (1964). 'On the Einstein-Podolski-Rosen Paradox'. *Physics*, **1**: 195–200.

Colbeck, R. and Renner, R. (2011). 'No Extension of Quantum Theory Can Have Improved Predictive Power'. *Nature Communications*, **2**: 411.

Colbeck, R. and Renner, R. (2012). 'Free Randomness Can be Amplified'. *Nature Physics*, **7**: 450–454.

Einstein, A. (1948). 'Quanten-Mechanik und Wirklichkeit'. *Dialectica*, **2**: 320–324. English translation: 'Quantum Mechanics and Reality'. Pp. 168–173 in M. Born (ed.), *The Born-Einstein Letters*. New York: Walker.

Gleason, A. N. (1957). 'Measures on the Closed Subspaces of Hilbert Space'. *Journal of Mathematics and Mechanics*, **6**: 885–893.

Heisenberg, W. (1925). 'Über Quantentheoretischer Umdeutung Kinematischer und Mechanischer Beziehungen'. *Zeitschrift für Physik*, **33**: 879–893.

Popescu, S. (2014). 'Nonlocality beyond Quantum Mechanics'. *Nature Physics*, **10**: 264–270.

Schrödinger, E. (1935). 'Discussion of Probability Relations Between Separated Systems'. *Proceedings of the Cambridge Philosophical Society*, **31**: 555–563.

Schumacher, B. (1995). 'Quantum Coding'. *Physical Review A*, **51**: 2738–2747.

Shannon, C. E. (1948). 'A Mathematical Theory of Communication'. *Bell System Technical Journal*, **27**: 379–423, 623–656.

Von Neumann, J. (1962). 'Quantum Logics (Strict- and Probability-Logics)'. Pp. 195–197 in J. von Neumann, *Collected Works*, Vol. 4. New York: Macmillan.

5

Quantum Information and Locality

DENNIS DIEKS

1 Introduction. What Makes Quantum Information Different: Non-orthogonality and Entanglement

The surprising aspects of quantum information are due to two distinctly non-classical features of the quantum world: first, different quantum states need not be orthogonal and, second, quantum states may be entangled. Of these two significant non-classical features non-orthogonality has to some extent become familiar and may therefore be called the less spectacular one; we shall only briefly review its consequences for information storage and transfer in this introductory section. By contrast, the conceptual implications of entanglement are amazing and still incompletely understood, in spite of several decades of debate. The greater part of this chapter (from Section 1 onward) is therefore devoted to an attempt to better understand the significance of entanglement, in particular for the basic physical concepts of "particle" and "localized physical system." It will turn out, so we shall argue, that the latter concepts have only limited applicability and that quantum mechanics is accordingly best seen as not belonging to the category of space-time theories, in which physical quantities can be defined as functions on space-time points. The resulting picture of the physical world is relevant for our understanding of the precise way in which quantum theory is non-local and sheds light on how quantum information processes can succeed in being more efficient than their classical counterparts.

Both in quantum and classical physics physical information is conveyed by messages that are encoded in states of physical systems; these states can be thought of as letters in an alphabet. Information transfer consists in the preparation of such messages by a sender and their subsequent reproduction (possibly in states of other systems) at the location of a receiver.

This characterization may seem to entail that the prospects of efficient information storage and transfer are generally worse in quantum theory than in classical physics. The reason is that in quantum theory different states of physical systems, even if maximally complete (pure states), may "overlap": they need not be represented by orthogonal vectors in Hilbert space. This non-orthogonality stands in the way of unequivocally distinguishing between states and so between the letters of the alphabet that is being used. Consider, for example, a message written in a binary code using states $|p\rangle$ and $|q\rangle$, with $\langle p|q\rangle \neq 0$; attempts to decode the message via measurements will inevitably lead to ambiguities. By contrast to the case of (ideal) measurements on classical states or on orthogonal quantum states, in which it is always possible to completely retrieve the original message, performing yes-no measurements represented by the projection operators $|p\rangle\langle p|$ and $|q\rangle\langle q|$ introduces a probability $|\langle p|q\rangle|^2$ of incorrect classification.

More sophisticated measurement schemes exist that do enable classification without error (Dieks 1988; Jaeger and Shimony 1995). However, the price to be paid is that in these schemes some measurements will have outcomes that lead to no classification at all. The best one can achieve this way is that a fraction $|\langle p|q\rangle|$ of all bits remains unclassified, so that non-orthogonality again leads to a loss of information.

The possibility of distinct though non-orthogonal states has the further consequence that statistical mixtures of quantum states cannot be uniquely decomposed into pure components. A statistical (impure) state ρ, prepared by probabilistically mixing an arbitrary number of non-orthogonal states $|\alpha_i\rangle$ with associated probabilities p_i, is the same mathematical entity as (and empirically indistinguishable from) the mixture of orthogonal states $|\lambda_i\rangle$ with probabilities λ_i, $i = 1, \ldots, n$, where the $|\lambda_i\rangle$ and λ_i are the eigenstates and eigenvalues of the density operator ρ ($\rho = \sum_i \lambda_i |\lambda_i\rangle\langle\lambda_i|$) and where n denotes the number of dimensions of the system's Hilbert space. From an informational viewpoint, the latter mixture (of at most n orthogonal and therefore perfectly distinguishable quantum states) is equivalent to a classical probability distribution with probabilities λ_i over n distinct classical states. Therefore, the amount of information that can be extracted from ρ is generally less than the (classical) information needed to specify the mixing procedure, in which the number of mixed states may be much larger than n.

This difference is made precise by the *noiseless quantum coding theorem* (Schumacher 1995). Essentially, this theorem tells us that the amount of information contained in a mixed state ρ is given by the *von Neumann entropy $S(\rho)$*, $S(\rho) = -Tr(\rho\log_2\rho)$. This von Neumann entropy of ρ is smaller than (or at most equal to) the classical *Shannon entropy* of the probability distribution p_i that determined the mixing process from which ρ originated. The latter entropy quantifies the amount of classical information needed to describe the mixing (where we take the different components $|\alpha_i\rangle$ as distinct classical possibilities). So instead of the classical result (Shannon 1948) that the Shannon entropy determines the number of bits needed to specify a state, we find that for quantum messaging the use of the smaller von Neumann entropy is appropriate.

This brief sketch illustrates how the *concept* of information does not change by the transition to quantum theory: as in the classical case, we are dealing with messages encoded in states of physical systems and with their transport through a physical channel. But the differences between classical systems and quantum systems, mirrored in the physical properties of their states, are responsible for differences in the *amounts* of information that can be stored and transferred.

The impression that quantum messaging is at best equally powerful as its classical predecessor would be incorrect, however. There are also situations in which quantum mechanics predicts more effective information transfer. The essential ingredient of these new quantum possibilities is the consideration of *composite* systems in *entangled states*. Entangled states represent systems that are correlated in a stronger way than deemed possible in classical physics, and enable instantaneous non-local influences between systems. This typical quantum non-locality underlies revolutionary new quantum information protocols like tele-portation. At least, this is the common view on the meaning of entanglement; we shall analyze it, and also criticize it, later.

2 Particle Individuality and Labels, Classical and Quantum

The conceptual landscape of classical physics is dominated by the notion of a localized individual system, with a *particle* as its paradigm case. Classical particles can be characterized by individuating sets of properties, like position, momentum, mass, and electric charge. Different classical particles may possess the same masses, charges, and momenta, but they cannot find themselves at the same position since repulsive forces are assumed to become increasingly strong when mutual distances become smaller. Consequently, two distinct particles differ in the values of at least one of their physical characteristics, namely their localization.

Within the framework of Newtonian physics, in which absolute space legiti-mates the notion of absolute position as a physical property, this guarantees that a version of Leibniz's principle of the identity of indistinguishable objects is respected: Newtonian particles are distinguishable from each other on the basis of their complete sets of physical properties and this distinguishability grounds their individuality. Within a relational space-time framework a weakened Leibnizean notion, namely that of weak discernibility, may be invoked to ground individuality (in cases like that of Black's spheres; see Dieks and Versteegh 2008 and references contained therein for further discussion). The general message is that in classical physics the specification, counting, and labeling of particles is done on the basis of physical distinctions, position being the most important.

Numerical particle labels $(1, 2, \ldots, N)$ accordingly correspond to such sets of physical properties; they have no content beyond this and function as conventionally chosen names. Thus, although it is true that a numerical labeling induces an ordering on sets of particles, this ordering is without physical significance and may be replaced by any other, permuted ordering: labels can be permuted in the equations without any physical consequence. It is therefore possible to switch over to a completely symmetrical description, in which all states that follow from each other by permutations of the particle labels are combined and the resulting complex is taken as the representation of *one* physical state (the "Z-star" introduced by Ehrenfest and Ehrenfest 1909). For example, in the simple case of two particles of the same kind (same mass, charge, and other kind-specific physical properties) – two electrons, say – we could have that one particle is at position $\vec{a}$ with momentum $\vec{\alpha}$ and the other at position $\vec{b}$, with momentum $\vec{\beta}$. This gives rise to two points in "phase space" (with axes p_1, q_1, p_2, q_2 representing the momenta and positions of the two particles, respectively): $p_1 = \vec{\alpha}$, $q_1 = \vec{a}$, $p_2 = \vec{\beta}$, $q_2 = \vec{b}$ and $p_2 = \vec{\alpha}$, $q_2 = \vec{a}, p_1 = \vec{\beta}, q_1 = \vec{b}$. But these two points correspond to the exact same physical situation, so that one is entitled to view the Z-star comprising *both* as representative of this single physical situation. This yields a description that is symmetrical in the particle labels and avoids an empirically empty "numerical identity" of the particles (alternatively, one could switch to a "reduced phase space" by "quotienting out" the equivalence generated by the $1 \leftrightarrow 2$ permutation).

The essential point is that classical systems are individuated by co-instantiated sets of physical properties and that labels possess the status of conventional names for such sets – they contain no information about the identity and individuality of particles beyond this. That physical properties are co-instantiated in fixed combinations is a basic principle of classical physics, with law-like status: physical properties only occur in specific bundles (like mass, position, and momentum, etc.) and not in isolation. These bundles define the physical individuals (Lombardi and Dieks 2016b).

Quantum mechanics *prima facie* respects a similar "bundle principle." Indeed, quantum systems like electrons are characterized by fixed sets of physical quantities: e.g., charge, mass, spin, and position. Compared to classical physics there is the complication that not all these quantum quantities can simultaneously be definite-valued, as a consequence of complementarity. (The characterization of quantum particles that we give here is not interpretation independent: e.g., in interpretations of the Bohm-type it is a first principle that the world consists of particles with always definite positions, contrary to what we assume here. Our analysis proceeds within a broadly construed no-hidden-variables approach.) Thus, position and momentum cannot possess sharp values simultaneously (although it is still possible to have both quantities in the same list of particle properties if one allows a certain unsharpness in their values – or "latitudes," as Bohr called it (Dieks 2017)). This incompatibility between sharp values of certain quantum quantities is taken into account through the use of non-commutative operator algebras instead of numerical quantities for the physical properties, plus the use of quantum states (vectors in Hilbert space) instead of classical states like (p, q). In this way we can generalize the classical idea of a set of physical properties whose values together define an individual system, by specifying an algebra of "observables" (Hermitian operators on a Hilbert space), together with a pure quantum state. In such a state the observable represented by the projection on the state, plus all observables commuting with this projection operator, are assigned definite values. This restricted value ascription is the best quantum mechanics can achieve: a complete set of quantum properties that takes the place of the classical property sets. Observables of which the state is not an eigenstate are characterized by probability distributions specifying the probabilities of possible outcomes in a measurement of the observables in question.

Modulo this complication of complementarity and non-definiteness in the numerical sense, the situation may seem not too different from the classical one: for example, instead of the values of position, momentum, and angular momentum characterizing a spinning classical particle we may now write down the corresponding position, momentum, and spin operators that characterize the system, e.g., an electron, and we can use a state like $|\psi\rangle|\rightarrow\rangle$ to characterize its space-time characteristics through $|\psi\rangle$ plus its spin through $|\rightarrow\rangle$.

Summing up, quantum particles can be characterized by fixed algebras of operators plus pure quantum states. (*Mixed* states do not characterize a single actual system when they are proper mixtures. Why improper mixtures, coming from entangled states, do not individuate particles is explained later.) The algebra, plus the state, gives us a complete set of quantum properties. The difference with classical mechanics thus appears to be solely in the replacement of definite numerical values of quantities with probability distributions determined by the quantum states. However, we shall see that things are not this easy, and that there are reasons to doubt the general applicability of the concept of a particle.

In the case of more than one particle, quantum theory works with labels that are meant to index the particles, in the same spirit as in classical physics. As before, we expect and require that labels only individuate by proxy, via physical properties or states. As we shall see, in the quantum case it is more difficult than before to put this simple principle into practice.

First, there is a technical complication due to the fact that the quantum mechanical state space is a vector space, in which the superposition principle holds.

The latter has the consequence that the natural way of forming symmetrical states is different from the Z-star procedure explained previously. Instead of the *collection* of permuted states, one now takes their *superposition*. In the case of particles of the same sort, this is precisely what the (anti-) symmetrization postulates demand: instead of considering product states of the form $|L\rangle_1|R\rangle_2$ (for a situation in which particle 1 is located to the left, particle 2 to the right) or $|R\rangle_1|L\rangle_2$ (the same physical situation, but with the particle labels permuted), we should write down (anti-) symmetrical states of the form

$$|\Psi\rangle = \frac{1}{\sqrt{2}}\{|L\rangle_1|R\rangle_2 \pm |R\rangle_1|L\rangle_2\}. \tag{1}$$

The state (1) is the quantum counterpart of Ehrenfest's symmetrical Z-star for two particles with the same intrinsic properties. If we look at the labels occurring in Eq.(1) we see that they are symmetrically distributed over the two terms in $|\Psi\rangle$ – so if we were to associate physical systems with these labels we would be forced to conclude that the two systems of (1) are in exactly the same state and therefore cannot be individuated by physical properties. The same conclusion follows in more technical fashion by taking "partial traces": tracing out over the parts of the total wave function labeled by 2 we obtain a mixed state for "system 1," namely $W = 1/2\{|L\rangle\langle L| + |R\rangle\langle R|\}$; and exactly the same state for "system 2" by tracing out over 1.

We have already observed in the classical case, however, that it would be wrong to uncritically rely on the referential power of labels: we should associate physical systems not with labels per se, but with physical properties. Just as the symmetry of the classical Z-star does not imply that no individual particles, each one with its own state, exist, the mere symmetry of $|\Psi\rangle$ in (1) does not compel us to abandon the notion of individual quantum particles. We should ask ourselves whether it is possible to individuate such particles by properties or states, rather than focusing on labels. In other words: we should be prepared to relax the one-to-one link between the labels that occur in the formalism (mathematically introduced as indices of the factor Hilbert spaces in the total tensor product space) and physical systems. The analogy with the classical case in fact suggests that the state $|\Psi\rangle$ of Eq.(1) corresponds to a situation with one particle located Left and one particle located Right, in particular if $|L\rangle$ and $|R\rangle$ stand for spatially widely separated wave packets that do not overlap.

Before embarking on the question of whether such an interpretation is justified, the following general remark is in order. The fact that a *superposition* of quantum states appears in Eq.(1), instead of a simple juxtaposition of possibilities, has an immediate non-classical consequence: the expectation value of an arbitrary operator P in the state $|\Psi\rangle$, $\langle\Psi|P|\Psi\rangle$ will contain an interference term $2\,\mathrm{Re}\langle LR|P|RL\rangle$. If this interference term is non-zero, i.e., when we are dealing with operators that have non-vanishing matrix elements between $|L\rangle$ and $|R\rangle$, and thus "connect" L and R, quantum mechanics predicts results for measurements of the physical quantity represented by P that are of the same type as the interference effects found in double-slit experiments. So even if an interpretation of (1) in terms of "one particle to the left and one particle to the right" is going to prove tenable, this will not mean a return to a *classical* particle picture: the systems that we are trying to identify are and remain quantum (Ladyman, Linnebo, and Bigaj 2013 discuss more details of the interference between the two "branches" in Eq.(1) and its empirical consequences in collision processes; they also point out that such interference terms can easily vanish, namely when there are additional physical features that distinguish between $|L\rangle$ and $|R\rangle$).

States of the type (1) are not product states, and this is usually taken to be the defining characteristic of entanglement. Furthermore, it is standard wisdom that identical particles in entangled states – electrons, say – do not possess pure states of their own, and that their mixed states are all the same. Therefore, physical individuation of component systems in such states seems impossible from the outset. But as we have noted, this standard account relies on the supposition that if there are individual systems, the *labels* occurring in the many-particles formalism will have to refer to them; and it is exactly this supposition that we are going to dispute.

To make the idea of the argument against the significance of "particle labels" clear, consider the EPR-Bohm state, notorious from the locality debates. The EPR experiment (in its modern Bohm version) is standardly discussed as being about two electrons at a large distance from each other, whose spins are correlated because the combination of their spins is described by the singlet state $1/\sqrt{2}\{|\uparrow\rangle_1|\downarrow\rangle_2 - |\downarrow\rangle_1|\uparrow\rangle_2\}$. The spatial part of the wave function is often not written down explicitly, but it obviously must be considered as well in order to make contact with the locality question. The total state including this spatial part has the form

$$|\Phi\rangle = \frac{1}{\sqrt{2}}\{|L\rangle_1|R\rangle_2 + |R\rangle_1|L\rangle_2\}\otimes\{|\uparrow\rangle_1|\downarrow\rangle_2 - |\downarrow\rangle_1|\uparrow\rangle_2\}, \qquad (2)$$

where $|L\rangle$ and $|R\rangle$ are states localized on the left and right, respectively, at a large distance from each other. Note that the spatial part of $|\Phi\rangle$ is symmetric in the "particle labels" 1 and 2, as required by the anti-symmetrization postulate (the total state must be antisymmetric, since we are dealing with fermions).

Now, if we use the state of Eq.(2) and also think that the labels 1 and 2 refer to our two particles, we have to conclude that there is neither a left nor a right electron.

The spatial states associated with both 1 and 2 are exactly the same, namely $1/2\{|L\rangle\langle L| + |R\rangle\langle R|\}$, so that the two particles defined this way would be "evenly distributed" between left and right. This means that the way the EPR case is standardly understood, as being about two localized particles at a large distance from each other, is at odds with the interpretation of the indices 1 and 2 as particle labels.

In the literature on EPR the state is sometimes given in a different form, namely as

$$|\Phi'\rangle = \frac{1}{\sqrt{2}}|L\rangle_1|R\rangle_2 \otimes \{|\uparrow\rangle_1|\downarrow\rangle_2 - |\downarrow\rangle_1|\uparrow\rangle_2\}, \tag{3}$$

in which the spatial part is a simple product state. This state is incorrect, because it violates the anti-symmetrization postulate – there are empirical differences between the states (2) and (3), because of the possibility of interference between the branches, as we have noted before. There can be no question that (2) is correct and (3) is not. But it is nevertheless understandable that (3) is sometimes written down, because this form of the state lends itself easily to the desired intuitive interpretation, namely that there is one single particle (particle 1) at the left, and another (particle 2) at the right. In Eq.(3) there is a one-to-one link between the labels and the physical location properties L and R, and this justifies the use of the labels as referring to the Left and Right regions, respectively. (A picture with particle 1 at L and particle 2 at R is still problematic, though: although the labels 1 and 2 correlate to positions via $|L\rangle$ and $|R\rangle$, respectively, there are no complete sets of quantum properties correlated to them. We shall discuss this issue extensively in the next section.)

However, for the sake of correctness we must work with the state (2) instead of (3). In this correct state there is no correspondence at all between individuating physical particle properties and the labels 1 and 2. This poses a dilemma: either we have to conclude that there are not localized systems in EPR-like situations, or we have to look for another way of referring to particles than by the indices occurring in the many-particles formalism.

Apparently, we have hit on an important difference between classical and quantum physics. In classical physics we can always individuate particles, or more generally physical objects, by unique sets of values of physical quantities. This justifies the use of particle labels, which serve as an abbreviation of such sets. By contrast, in quantum theory it is at least unclear whether objects can be referred to in this way – this might already raise doubt about the general appropriateness of concepts like "particle" and "localized object" in the quantum context.

3 Identical Quantum Particles

If it were the case that the "particle labels" in an EPR-like situation constituted our only basis for speaking about physical systems, we would be forced to accept (among other things) that these systems are not localized in small regions in space: they would be "half in L and half in R." A measurement on the left wing of the EPR experiment could in this case perhaps equally affect particle 1 and particle 2. This might be considered to possess a positive side: it could perhaps be taken as the starting point for an explanation of the non-local quantum effects that lie at the basis of violations of Bell inequalities, and also of the effectiveness of quantum information processes. If the physical systems that are involved in such processes are non-local to start with, non-locality in their behavior should hardly surprise us. But the price to pay for attempts to use this strategy in this exact form would be high. We would be compelled to accept that all identical particles, electrons, for example, would be smeared out over the universe. And, importantly, we would be unable to recover the classical particle picture in the classical limit of quantum theory. In the classical limit it should surely be possible to think in terms of localized particles that can be labeled on the basis of their positions and trajectories – but this means that these classical labels cannot be in one-to-one correspondence with the indices occurring in the quantum formalism (Dieks and Lubberdink 2011). So there must be a different and better way of identifying quantum particles if we wish to have a smooth transition from quantum to classical. When we have found how to do this, we should look at the explanation of non-local effects afresh.

Fortunately, an alternative way of identifying quantum systems is not difficult to find. The crux is in our earlier observation that particle labels have no physical content in themselves and only obtain such content via association with physical quantities. These physical quantities and physical states are consequently primary; the labels are secondary auxiliary tools. As Ghirardi, Marinatto, and Weber (2002) point out (see also the discussion by Ladyman et al. 2013 and by Caulton 2014), in the case of particles of the same kind (identical particles), it is accordingly necessary to allow only physical quantities that are symmetrical in the labels. Indeed, think back to Ehrenfest's Z-star for two classical particles: there, we should count $\vec{a}_1, \vec{b}_2$ and $\vec{a}_2, \vec{b}_1$ as *one* physical situation, which will automatically happen if we symmetrize the physical quantities with which we characterize the situation. Analogously, instead of considering particle properties represented by projection operators of the form $P_1 \otimes I_2$, with I_2 the unity operator in Hilbert space 2, we should in the quantum case focus on projection operators like

$$P_1 \otimes I_2 + I_1 \otimes P_2 + P_1 \otimes P_2. \tag{4}$$

The expectation value of the projector in (4), in a properly (anti-)symmetrized state, gives the probability of finding at least one of the identical particles in the state onto which P projects. The last term in (4) can be left out without consequences in the case of fermions; it has been added to allow for the possibility that the two particles are found in exactly the same state, which could occur when we are dealing with bosons. Let us stress that the analogy with the classical case is not the only reason for symmetrization of the observables: consistency of the formalism requires it as well. For example, a non-symmetrical Hamiltonian would steer the state out of its subspace of (anti-)symmetrical states and would thus create a conflict with the (anti-)symmetrization postulate.

As Ghirardi, Marinatto, and Weber (2002) observe, it is in this way possible to assign a complete set of quantum properties to components of a multi-particles system, even in symmetrical or anti-symmetrical states, if the following condition is satisfied: the total (entangled) state should be obtained by symmetrizing or anti-symmetrizing a *factorized* state. Such states are eigenvectors of symmetrical projection operators of the type (4). Making use of these sets of quantum properties, we may identify subsystems via their *states* instead of via labels.

This identification of subsystems is in line with the laboratory practice of physics; it also respects the classical limit (Dieks and Lubberdink 2011; Dieks 2014). To see the essential idea in a simple example, consider the case of two fermions in state

$$\frac{1}{\sqrt{2}}\{|L\rangle_1|R\rangle_2 - |R\rangle_1|L\rangle_2\}, \tag{5}$$

with $|L\rangle$ and $|R\rangle$ two non-overlapping wavefunctions at a large distance from each other. The property assignment by means of the projection operators $|L\rangle_1\langle L|_1\otimes I_2 + I_1\otimes|L\rangle_2\langle L|_2$ and $|R\rangle_1\langle R|_1\otimes I_2 + I_1\otimes|R\rangle_2\langle R|_2$ leads to the conclusion that we have one particle characterized by $|L\rangle$ and one particle characterized by $|R\rangle$, respectively (we have left out the last term of (4) in the projection operators, as this term automatically vanishes in the case of fermions). The state (5) represents therefore one particle at the left and one at the right, in spite of the fact that the labels 1 and 2 are evenly distributed over L and R. In the classical limit, this would correspond exactly to what we expect: particles, e.g., electrons, represented by narrow wave packets, following approximately classical trajectories (narrow wave packets spread out, and will therefore only be able to follow approximately classical trajectories for a limited time; for the classical limit description to be applicable, additional conditions must be fulfilled, like the presence of decohering mechanisms; see Rosaler 2016). These classical objects can of course be labeled, on the basis of their unique positions and trajectories, but the labels thus defined will not coincide with the indices 1 and 2 occurring in the quantum state – the latter remain evenly distributed over the wave packets even in the classical limit.

We have glossed over an important point, namely that the decomposition of states like (1) generally is not unique, because of the equality of the coefficients appearing in front of the terms in the superposition ("degeneracy"). So alternative descriptions, in addition to the one in terms of $|L\rangle$ and $|R\rangle$, are possible. To make the description unique the symmetry needs to be broken; it is plausible that such a symmetry breaking will be realized when the two-particle system enters into interaction with a macroscopic environment, via position-dependent interactions. This issue is important for non-collapse interpretations of quantum mechanics, like the many-worlds interpretation or the modal interpretation (see Lombardi and Dieks 2016a and references therein). In the present context, it is sufficient to focus on the consistency of the property assignment in terms of L and R.

The states obtained by (anti-)symmetrizing product states can thus be understood as representing particles possessing their own independent quantum properties. In this sense they are *not* entangled, in spite of the fact that the total state is not factorizable. This may seem strange since the singlet state, the paradigm example of entanglement, appears to be precisely such an anti-symmetrized product. However, the complete state in Bell-type experiments is of the form (2) rather than just $\{|\uparrow\rangle_1|\downarrow\rangle_2 - |\downarrow\rangle_1|\uparrow\rangle_2\}$. This complete state is *not* the result of anti-symmetrizing a product state, even though its spin part (the singlet) is. This property of the total state is responsible for the non-factorizability of joint probabilities for measurement outcomes on the two wings of the Bell experiment, and consequently for violations of the Bell inequality and for non-locality. By contrast, had we begun with a factorizable state like $|L\rangle_1|R\rangle_2|\uparrow\rangle_1|\downarrow\rangle_2$, we would after anti-symmetrizing have arrived at

$$\frac{1}{\sqrt{2}}\{|L\rangle_1|R\rangle_2|\uparrow\rangle_1|\downarrow\rangle_2 - |R\rangle_1|L\rangle_2|\downarrow\rangle_1|\uparrow\rangle_2\}. \tag{6}$$

The probabilities in this state, for spin measurements on the two wings of the experiment, *do* factorize. This means that there will be no violations of Bell inequalities and no no-go results for local models. In fact, the quantum formalism itself, with the aforementioned provisos about the meaning of labels, immediately suggests a local model: the state (6) describes a situation in which there are two particles, one left and one right, with the particle on the left-hand side having its spin up and the right-hand particle having its spin down.

This brings us to the essential difference between the states (2) and (6). In (6) there is a strict correlation between spatial and spin properties, which is absent in (2). In terms of the particle concept that we have previously discussed, we can say that (6) makes it possible to think of two particles labeled by $(L, \uparrow)$ and $(R, \downarrow)$, respectively (note that these differ from the standard labels 1 and 2). In (2) there is no such correlation, and this means that we cannot define physically meaningful particle labels that stand for co-instantiated sets of properties. In other words, (6) lends itself to a particle interpretation but (2) does not.

4 EPR without Particles

The upshot of the discussion so far is that although a focus on particle labels makes it initially appear that in quantum mechanics there is no place at all for localized particles of the same kind, reflection on the meaning of the labels redresses the balance. Proper attention to what we *mean* by "particles" makes clear that we *can* have situations of individual, even though "identical" particles, each with its own distinctive properties. We can therefore resist the idea that quantum particles are essentially non-local entities: it certainly is possible to have two identical quantum particles, one localized at $|L\rangle$ and the other at $|R\rangle$.

However, further reflection shows that exactly those situations in which we can define quantum particles by specifying sets of individuating properties, are situations that are uninteresting from a quantum information point of view. For example, the state (6) allows the definition and individuation of two particles, each characterized by its own position and spin. It is a state that respects the traditional particle concept, by tying properties together in definite bundles, but it does so at the cost of losing its quantum interest. Indeed, as we have seen, joint two-particle probabilities factorize in (6), so this state satisfies all locality requirements and will not violate any Bell inequalities. It will consequently not be suited for informational purposes that transcend classical possibilities.

By contrast, the state (2) defies attempts at defining particles by complete sets of properties: there is no correlation, in this state, between positions and spins. The very concept of a particle becomes moot in this case. Although it may still seem natural to treat L and R as a kind of "particle labels" here, since we can only measure something at these positions, these labels do not correlate to other properties than position itself. This strange situation is the one that is vital from a quantum information viewpoint, but at the same time it is a situation in which the classical particle concept breaks down.

It might be felt that this is just what was to be expected, since the violation of Bell inequalities is commonly acknowledged to demonstrate that no local account can be given of what happens in an EPR experiment – we could have anticipated from this that we should not speak of one particle at position L and another at R, each with its own properties. But this is not the moral that is usually drawn. The standard story is instead that there *are* two localized particles, but that a measurement on particle 1 instantaneously *changes* some feature of particle 2. This story presupposes the traditional concept of a particle with its own individual properties. Our analysis goes against this standard story by arguing that the traditional conceptual framework is inadequate.

Looking at (2), we see that in this state there is no correlation between spins and positions; the total state is the *product* of its spatial and its spin parts. So even if L is taken as a label (instead of 1 or 2), we are not allowed to refer to the "spin of the particle on the left." Instead, there is a double-spin state that combines equally with both L and R. It might seem that on balance this returns us to a picture of the kind we have alluded to before, in which particles are non-local (present both at L and R) from the very start. But it would be inaccurate to make no distinction between what we are arguing and proposing here and that earlier suggestion of non-locality.

What we briefly considered earlier was the possibility that there are two particles, but that both are non-local entities in the sense that they are present at L *and* R. But what we argue for here is the *inadequacy* of the particle concept in typical quantum cases. According to this new viewpoint, there are only two positions in space (L and R) where something can be measured; and in both places the *very same* two-spin state is "visible." This spin state is not defined as a function on spatial points, so it is not a physical *field* (a field, e.g., an electric or magnetic field, is an assignment of field strengths to spatial points; the values of the field are therefore correlated to positions, as is clear from the notation $\vec{E}(x)$). By contrast, the spin state lives in its own Hilbert space, independently of the spatial part of the wave function. In other words, we should not think of spin amplitudes as being non-zero at L and R and zero at other positions, in the same way as in the case of a classical electric field that is only non-vanishing at L and R. The spins by themselves do not constitute a spatial entity at all.

Classical fields could have the same *numerical values* for their field strengths at L and R, but these field strengths would obviously not be identical in the metaphysical sense (according to which two identical things are one and the same). The proposal for the quantum case that we are considering is very different: at both L and R we can make contact with the *identically same* spin state – it is as if both L and R are windows through which we are able to look at exactly the same scene (which itself is not spatial). As it turns out, this picture leads to an explanation of the EPR-Bohm experiment that may be called "local," even though this explanation is essentially non-classical and dispenses with the particle picture.

5 Local Quantum Explanations

Let us look at the details of what happens if an EPR-Bohm experiment is performed in the state (2). What usually is taken to be the phenomenon in need of explanation is that a spin measurement on the left wing of the experiment produces a definite spin value on the right wing, while there was no such definite value before and in spite of the fact that no signal propagates from one side of the experiment to the other. As we have argued, this way of formulating the problem presupposes that there is a correlation between spin values and positions – that we have particles at the two wings of the experiment that possess their own spin characteristics (non-definiteness of the spin value being such a spin characteristic), which are subsequently changed by the measurement. Our rejection of this presupposition thus dissolves the problem as originally posed. But we should of course investigate the explanatory resources of our own alternative account to see whether this change of perspective is of any help.

Let us suppose that a spin measurement in the z-direction is made by a device located in the region L, and that the result is "up." Let us first consider what happens if we make use of the projection postulate (collapse of the wavefunction); this is the treatment that in the usual approaches most strongly suggests that a superluminal influence propagates between the wings of the experiment. Application of the projection postulate within our conceptual framework leads to the following account. The spin-measuring part of the device is able to make contact with the singlet state (the position-independent double spin state) via the "space window" located at L. The outcome "up" gets registered by the device at L, and as a result the singlet state collapses to the spin-down state $|\downarrow\rangle$. *This very same state* is subsequently observed from the "window" R when a spin measurement is undertaken at that position. There is no need for any signal between L and R, since the spin state is not spatial. The spin state is identically the same regardless of whether seen from L or R.

Formally, the treatment with the projection postulate goes as follows. The projection operator that represents the spin measurement on the left wing is

$$P = \left\{ |L\rangle_1 \langle L|_1 |\uparrow\rangle_1 \langle\uparrow|_1 \otimes I_2 + I_1 \otimes |L\rangle_2 \langle L|_2 |\uparrow\rangle_2 \langle\uparrow|_2 \right\}, \tag{7}$$

where account has been taken of the fact that the spin measurement is at the same time a position measurement, by a localized device. The projectors representing the spin measurement are therefore correlated to position. The result of making P work on $|\Phi\rangle$ of Eq.(2) is

$$\frac{1}{\sqrt{2}} \left\{ |L\rangle_1 |R\rangle_2 |\uparrow\rangle_1 |\downarrow\rangle_2 - |R\rangle_1 |L\rangle_2 |\downarrow\rangle_1 |\uparrow\rangle_2 \right\}. \tag{8}$$

which is a properly anti-symmetrized product state. As we have seen in Section 3, such product states represent particles in the usual sense, for which spatial properties and spin properties are bound together in an identifying set of properties.

The explanatory story remains essentially the same as just sketched: the physical interaction localized at L transforms the singlet state to $|\uparrow\rangle_L|\downarrow\rangle_R$. Before the measurement, the spin state was not a spatial entity (it was not correlated to the spatial wavefunction), but the local measurement at L correlates "up spin" to $|L\rangle$. The "down spin" becomes consequently correlated to $|R\rangle$, not because of a signal going from L to R, but because the event of establishing the latter correlation is identical to the measurement at L. In this sense the explanation is *local*.

Effectively the same story can also be told without collapses, employing only unitary time evolution. This is the account that should be used in order to give a local explanation of EPR-type experiments within non-collapse interpretations of quantum mechanics. In the familiar schematic von Neumann rendering, in which a measuring device possesses a neutral state $|M_0\rangle$ and post-measurement states $|M_\uparrow\rangle$ and $|M_\downarrow\rangle$, we obtain after a measurement interaction in which only local interactions are involved:

$$|\Phi'\rangle = \frac{1}{\sqrt{2}}\left\{ \begin{array}{l} |M_\uparrow\rangle(|L\rangle_1|R\rangle_2|\uparrow\rangle_1|\downarrow\rangle_2 - |R\rangle_1|L\rangle_2|\downarrow\rangle_1|\uparrow\rangle_2) + \\ |M_\downarrow\rangle(|R\rangle_1|L\rangle_2|\uparrow\rangle_1|\downarrow\rangle_2 - |L\rangle_1|R\rangle_2|\downarrow\rangle_1|\uparrow\rangle_2) \end{array} \right\}. \tag{9}$$

So, relative to the device state $|M_\uparrow\rangle$ representing the outcome "up" we get a properly anti-symmetrized product state according to which there is a left-hand particle with spin up and a right-hand particle with spin down. This does not require any non-local interactions precisely because the spin part of the original state $|\Phi\rangle$, i.e., the singlet state, is uncorrelated to position. The local spin measurement at L in effect (if we think of the anti-symmetrized product state as an effective product) selects from this singlet the part $|\uparrow\rangle|\downarrow\rangle$, and this is possible via an interaction that is restricted to the region L because the whole singlet is available there.

The mechanisms that lie at the basis of quantum information transfer can thus be understood in a local way, provided that we realize that the classical notion of a particle, or a field of properties defined on points in space, is not generally applicable in quantum theory.

6 From Classical to Quantum

In Section 5, we considered examples of transitions from a general quantum picture, in which there are no sets of interconnected properties defining particles (or sets of properties tied to a spatial point), to a classical picture in which there *are* such particle or field properties. The existence of these transitions shows that quantum theory is able to accommodate classical particles and fields as special cases, as we would expect from a theory that is more general than classical physics. But the opposite transition should also be possible: starting from a classical situation of localized systems, with their own properties, it should be feasible to produce typical quantum situations. If that were not the case, it would be impossible to design laboratory experiments in which quantum teleportation and other entanglement-assisted quantum information transfer processes manifest themselves.

The simplest example of such a transition from classical to quantum is the paradigmatic Bell case, in which a localized system, initially with its own individuating properties (e.g., total spin $S = 0$), decays into two parts that do not behave as particles, in the sense that they do not possess individuating properties (e.g., the definite-valued spin state $S = 0$ splits into two parts whose combined spins are described by the singlet state). Suppose that the two spatial wave packets subsequently travel far apart due to free propagation. The spin part does not change by this since it is not spatial and unrelated to the evolution responsible for the spatial traveling. This finally leads to the EPR-Bohm state, which is not a particle state as we have seen, even though it has two spatial parts at a distance. The transition from classical to quantum took place locally here, in the decaying process, and was unaffected by the subsequent traveling of the spatial parts.

A more sophisticated experiment with entangled electrons has recently been discussed and implemented by Hensen and colleagues (2015); compare also the alternative experiment with photons by Giustina, Versteegh, and colleagues (2015). The results of these experiments are important for various reasons, not the least of which is that they hold prospects for applications in quantum information protocols. In the Hensen experiment, remote electron spin entanglement is created without traveling electrons (following a proposal by Barrett and Kok 2005). At the start of this experiment two localized electrons, both in definite spin states and therefore comparable to classical particles, are prepared at stable positions A and B, at a large distance (1.3 km) from each other. The two initial spin states are the same, namely $|s\rangle = 1/\sqrt{2}(|\uparrow\rangle + |\downarrow\rangle)$, so that the two-electron state at this stage can be represented by

$$\frac{1}{\sqrt{2}}\{|\phi_A\rangle_1|\phi_B\rangle_2 - |\phi_B\rangle_1|\phi_A\rangle_2\}|s\rangle_1|s\rangle_2. \tag{10}$$

Note that it would make no sense to say that in this state one particle finds itself at A and the other at B if we assumed that the labels 1 and 2 refer to particles. As we observed before, the standard terminology in experimental physics presupposes an analysis of the individuality of particles as we have given in Section 3.

In the next step of the process the electrons at the two positions are excited by a short laser pulse, after which spontaneous photon emission entangles the local electrons spins with the photon field in such a way that the spins together with the field are described at both wings by the state $1/\sqrt{2}(|{\uparrow}1\rangle + |{\downarrow}0\rangle)$, in which 1 and 0 denote the presence and absence of a photon, respectively. The total state is accordingly transformed from (10) into

$$\frac{1}{\sqrt{2}}\Big\{|\phi_A\rangle_1|\phi_B\rangle_2 - |\phi_B\rangle_1|\phi_A\rangle_2\Big\}\otimes 1/\sqrt{2}\{|{\uparrow}1\rangle_1|{\uparrow}1\rangle_2 + |{\downarrow}0\rangle_1|{\downarrow}0\rangle_2$$

$$+|{\uparrow}1\rangle_1|{\downarrow}0\rangle_2 + |{\downarrow}0\rangle_1|{\uparrow}1\rangle_2\Big\}. \tag{11}$$

To simplify the notation we have indexed the photon/spin states with the associated electron spin labels.

Now, if a one-photon state is detected at a third position, possibly very distant from A and B, without any discrimination with respect to its provenance from either A or B, the state (11) effectively collapses to its part containing one photon ($|10\rangle$ or $|01\rangle$):

$$\frac{1}{\sqrt{2}}\Big\{|\phi_A\rangle_1|\phi_B\rangle_2 - |\phi_B\rangle_1|\phi_A\rangle_2\Big\}\otimes 1/\sqrt{2}\Big\{|{\uparrow}\rangle_1|{\downarrow}\rangle_2 + |{\downarrow}\rangle_1|{\uparrow}\rangle_2\Big\}. \tag{12}$$

This is a maximally entangled spin state, comparable to the EPR-Bohm state of Eq.(2).

The actual procedure followed in the experiment by Hensen and colleagues (2015) is more sophisticated and involves a further step to eliminate the "detection loophole," which in this case is the possibility that only one photon is detected, due to incomplete detector efficiency, whereas in reality two were emitted. This additional step guarantees that detection of a photon "heralds," with certainty, the existence of an entangled pair of electron spins, associated with the localized spatial electron wave functions at A and B.

In this experimental procedure the initial state was classical in the sense that it described two well-localized electrons, each with a definite spin. This familiar picture became invalid when the electrons interacted with the electromagnetic field: through purely local interactions (resulting in spontaneous photon emissions) the electron spins were entangled with the photon field. After this entanglement the electron positions A and B were no longer associated with pure spin states, so that the classical particle picture ceased to be applicable: although electron position measurements could still only have success at A and B, it was no longer true that electrons with complete sets of electron properties were present at these positions. The (again purely local) measurement on the photon field, which may have taken place at a large distance from both A and B, finally fully decoupled the spins from the positions A and B so that the total state (12) became a product of spatial and spin parts.

7 Conclusion

The classical theories of physics, namely, non-quantum theories up to and includ-ing general relativity, are all *space-time theories* in the sense that they define physical systems through sets of properties that are functions on a manifold of space-time points. This theoretical feature immediately implies a *basic principle of locality* that is satisfied, namely that physical systems and their histories can be completely described by the specification of all the local states of affairs in their existence. A paradigmatic example of such a classical system is a particle, in particular a point-particle: all its properties, including its position, are bound together in one package, which defines the particle and its state. This first locality principle has to be distinguished from a second one, namely that signals cannot travel faster than light. When we combine the two principles we arrive at the result that a particle at R cannot undergo any changes due to what happens at L, unless there is enough time for a causal signal to propagate from L to R.

The structure of the formalism of quantum mechanics suggests an entirely different picture. At first sight, it may seem that in the case of particles of the same kind ("identical particles") there is no scope for the concept of a localized particle at all, because of the (anti-)symmetrization rules: the particle labels are evenly distributed over all positions where the total many-particles wavefunction does not vanish. But this first obstacle for a particle interpretation can be quickly removed, by individuating the particles via their states instead of via the labels that occur in the formalism. However, this maneuver has only limited success, because it leads only to an analog of the classical particle picture in quite specific circum-stances. In the general quantum situation, "particle properties" do not combine to form individual particle states. In particular, it is possible to have localization in individual narrow regions in the spatial part of the total state, without complete packages of particle properties that are correlated to these individual regions. The basic locality principle just mentioned ("locality in the first sense") therefore fails in quantum mechanics.

This does not entail that the second locality principle that we mentioned, namely, the impossibility of superluminal signaling, fails too. This second principle is respected in quantum mechanics, but this is only relevant for the explanation of correlations between measurement results if we are dealing with a space-time theory in which properties are local in the sense of our basic notion of locality. This means that it is true that quantum features that are bound to position cannot be changed except by signals that have enough time to reach them, just as in classical physics. But according to the quantum formalism, there is also the possibility of physical properties that are independent of space, like the spins of the singlet state. When such space-independent properties are modified by some intervention, the effect of the modification is valid for the total system at once. Such a change suggests an action at a distance when interpreted as a change in *local* features; but our point is precisely that in these cases we should *not* think in terms of local properties, not even when the spatial part of the total wavefunction is restricted to well-defined narrow regions.

The new possibilities of quantum information transfer depend on this lack of *basic locality* in quantum mechanics; they do not conflict with locality in the second sense and do not conflict with relativity theory. Awareness of this double role of locality and the precise way in which quantum mechanics is non-local appears to provide a framework for understanding the effectiveness of quantum information processes.

References

Barrett, S. D. and Kok, P. (2005). "Efficient High-Fidelity Quantum Computation Using Matter Qubits and Linear Optics." *Physical Review A*, **71**: 060310.

Caulton, A. (2014). "Qualitative Individuation in Permutation-Invariant Quantum Mechanics." arXiv:1409.0247v1 [quant-ph].

Dieks, D. (1988). "Overlap and Distinguishability of Quantum States." *Physics Letters A*, **126**: 303–306.

Dieks, D. (2014). "The Logic of Identity: Distinguishability and Indistinguishability in Classical and Quantum Physics." *Foundations of Physics*, **44**: 1302–1316.

Dieks, D. (2017). "Niels Bohr and the Formalism of Quantum Mechanics." Forthcoming in H. Folse and J. Faye (eds.), *Niels Bohr and Philosophy of Physics: Twenty First Century Perspectives*. London: Bloomsbury Publishing.

Dieks, D. and Lubberdink, A. (2011). "How Classical Particles Emerge from the Quantum World." *Foundations of Physics*, **41**: 1051–1064.

Dieks, D. and Versteegh, M. A. M. (2008). "Identical Particles and Weak Discernibility." *Foundations of Physics*, **38**: 923–934.

Ehrenfest, T. and Ehrenfest, T. (1909). *Begriffliche Grundlagen der statistischen Auffassung in der Mechanik*. Leipzig: Teubner. English translation: *Conceptual Foundations of the Statistical Approach in Mechanics*. Ithaca, NY: Cornell University Press, 1959.

Ghirardi, G., Marinatto, L., and Weber, T. (2002). "Entanglement and Properties of Composite Quantum Systems: A Conceptual and Mathematical Analysis." *Journal of Statistical Physics*, **108**: 49–122.

Giustina, M., Versteegh, A. M. et al. (2015). "Significant-Loophole-Free Test of Bell's Theorem with Entangled Photons." *Physical Review Letters*, **115**: 250401.

Hensen, B. et al. (2015). "Loophole-Free Bell Inequality Violation Using Electron Spins Separated by 1.3 Kilometers." *Nature*, **526**: 682–686.

Jaeger, G. and Shimony, A. (1995). "Optimal Distinction between Two Non-orthogonal Quantum States." *Physics Letters A*, **197**: 83–87.

Ladyman, J., Linnebo, Ø., and Bigaj, T. (2013). "Entanglement and Non-factorizability." *Studies in History and Philosophy of Modern Physics*, **44**: 215–221.

Lombardi, O. and Dieks, D. (2016a). "Modal Interpretations of Quantum Mechanics." In E. N. Zalta (ed.), *The Stanford Encyclopedia of Philosophy*, Winter 2016 Edition. https://plato.stanford.edu/archives/win2016/entries/qm-modal/

Lombardi, O. and Dieks, D. (2016b). "Particles in a Quantum Ontology of Properties." Pp. 123–143 in T. Bigaj and C. Wüthrich (eds.), *Metaphysics in Contemporary Physics*. Leiden: Rodopi-Brill.

Rosaler, J. (2016). "Interpretation Neutrality in the Classical Domain of Quantum Theory." *Studies in History and Philosophy of Modern Physics*, **53**: 54–72.

Schumacher, B. (1995). "Quantum Coding." *Physical Review A*, **51**: 2738–2747.

Shannon, C. E. (1948). "The Mathematical Theory of Communication." *Bell Systems Technical Journal*, **27**: 379–423, 623–656.

6

Pragmatic Information in Quantum Mechanics

JUAN ROEDERER

1 What Is Pragmatic Information?

We live in the information era. It is quite understandable that a term of such everyday usage as "information" remains largely undefined when used in the general scientific literature, despite the fact that philosophers, mathematicians, linguists, engineers, biologists, and writers may use this common term with quite different and distinct meanings. Physicists are accustomed to working with the mathematically well-founded concepts of Shannon (or statistical) and algorithmic information. These terms are used to designate objective, quantitative expressions of the *amount* of information in a message, the *gain* of information when alternatives are resolved, the *degree* of uncertainty, the *quality* of transmitted signals, the minimum *number* of binary steps to identify or describe something, the maximum number of bits that may be processed in one location, the total number of bits available in the Universe, etc. However, whenever physicists use the term "information" in a more general way implying the notions of meaning or purpose, such as in statements like "the field carries information about its sources," "information cannot travel faster than light," "the DNA molecule carries information about the organism," "information about the microstate of a gas," "the detector provides information about the radiation background," "information about the initial state of a system," "information deposited in (or extracted from) the environment," etc., they are really talking about *pragmatic information.*

In effect, it is mostly the biologists, particularly geneticists and neuroscientists, who make an explicit distinction between Shannon or statistical information (e.g., number of bits to transmit a given message or to resolve some alternatives, detached from purpose or effect), algorithmic information (minimum number of binary steps to define something; relation between values of physical variables given by an equation), and pragmatic information (involving purpose, cause and effect) (for a comprehensive discussion, see Küppers 1990). Of these three classes of information, the last one is usually the most relevant in biology; indeed, the notion of *quantity* of information is often of secondary importance: what counts is what information ultimately *does*, not how many bits are involved.

As I argue elsewhere (Roederer 2003, 2005, 2016), pragmatic information is a purely biological concept. This seems quite difficult to accept for physicists, not the least because many agree with John Wheeler's dictum that "every physical quantity derives its ultimate significance from bits, binary yes-or-no indications" (1989). Yet let me emphasize the trivial fact that physics is the product of information processing by *brains*, based on interactions of systems chosen and prepared by *brains*, measured with artifacts (instruments) created by *brains*, using algorithms (mathematics and models) developed by *brains*, and following plans, purposes, and quantitative predictions made by *brains*. Perhaps the principal obstacle in an effort to persuade physicists to accept the idea of information being a purely biological concept is the perceived lack of a widely accepted definition of pragmatic information that is truly objective, i.e., unrelated to any human use or practice (from its beginning, traditional information theory – Shannon and Weaver 1949 – deliberately refrained from giving a universal and objective definition of the concept of information per se).

Philosophers have struggled for a long time with an objective definition of information. Physicists often consider information in its general sense to be a statement that resolves a set of alternatives, i.e., which resolves uncertainty. Others consider "data" to be the essence of information. Most visualize information as a quantity expressed in number of bits. Rather than attempting to define information ab initio, I find it more appropriate to start with *interaction between bodies* as the primary concept or "epistemological primitive" (Roederer 1978, 2005), and from there derive the concept of information. We can identify two distinct groups. Category 1) interactions can always be reduced to a linear *superposition* of physical interactions (i.e., forces) between the interacting systems' elementary constituents, in which energy transfer between the interacting parts plays a fundamental role. Category 2) interactions cannot be expressed as a superposition of elementary interactions, and it is *patterns and forms* (in space and/or time) that play the determining role in whether an interaction is to take place; the required energy must be provided from outside through a specific complex *interaction mechanism*.

Examples of category 1), which we call "force-driven" interactions, are all the physical interactions between elementary particles, wave fields, nuclei, atoms, molecules, parcels of fluid, complex solid bodies and networks, planets, and stars. Ultimately, they all originate in the four basic interactions between fundamental particles (electromagnetic, gravitational, strong, and weak). It is the fundamental property of reducibility of force-driven interactions that allows physics to work with approximate *models* of the complex reality "outside" to make quantitative predictions or retrodictions about the time-evolution of a given physical system.

The simplest case of an interaction of category 2) is any arrangement in which the presence of a specific pattern in a complex system S (sometimes called "the sender" or "the source") leads to a causal, macroscopic, and *univocal* change in another complex system R (the recipient), a change that would not happen (or just occur by chance) in the absence of the particular pattern at the source. Typical examples range from effects on their respective chemical environments of the one-dimensional pattern of bases in the RNA molecule or the three-dimensional shape of a folded protein; the light patterns detected by an insect and the resulting shape of its flight orbit around a light source; the patterns of neural electrical impulses in one region of the brain triggering impulses in another; to the perceived print patterns changing the state of knowledge in a reader's brain. It is important to point out that information-driven interactions all require a complex *interaction mechanism* with a reset function (often considered part of one of the interacting bodies), and which ultimately provides the energy required to effect the specific change. Although we call them "interactions," they are unidirectional, representing a cause-and-effect relationship between source and recipient. However, the designation *inter*action is justified in the sense that, to occur, they require some predesigned or evolutionary match (sometimes called "understanding") between source and recipient – precisely, the interaction mechanism.

Pragmatic information is then defined as *that which represents the univocal correspondence pattern $\rightarrow$ change*; it is the reason why I call Category 2 "information-driven interactions" (Roederer 2003). This may appear to some readers as a circular definition; it is not: "information-driven" is just a convenient name for this category; I could have called them "pattern-driven interactions." By "univocal" I mean that the interaction process is deterministic and must yield identical results when repeated under similar conditions of preparation (I shall ignore at this stage considerations of matching errors, fluctuations, etc.). This in turn means that the triggering pattern must be stable during some finite time and be amenable to be copied. When an information-driven interaction has occurred, I say that pragmatic information was transferred from the source S (where the pattern resides) to the recipient R (where the pattern-specific change occurs). I wish to emphasize that, in this definition, the concept of pattern refers to any physical/geometric/topological property of points in space and/or instants of time that distinguishes them from all others (in philosophy, there is a large literature on *semiotics*, the study of signs, their meaning, and effects, first developed by C. S. Peirce in the late nineteenth century); this includes, but is not limited to, symbols to which one can assign syntactic and semantic dimensions (e.g., see Küppers 1990: chapter 3).

Note that in the preceding we do not imply that information "resides" in the patterns – the concept of pragmatic information is one of *relationship* between patterns and changes, mediated by some interaction mechanism. A pattern all by itself has no meaning or function. On occasion one may say that information is *encoded* in the pattern at the source. The effected change can itself be a new pattern. Finally, there is no such thing as a numerical *measure* of pragmatic information. Pragmatic information cannot be quantified – it represents *a correspondence* that either exists or not, or works as intended or not, but it cannot be assigned a magnitude. There are cases, of course, where a given pattern and/or the change can be expressed in numbers or bits by an external agent, but that number *is not* the pragmatic information involved.

There are only three fundamental processes through which mechanisms of information-driven interactions can emerge (Roederer 2005), involving processes at three vastly different time scales: 1) Darwinian evolution; 2) adaptation or neural learning; 3) the result of human reasoning and long-term planning. In other words, they all involve *living matter* – indeed, information-driven interactions represent the *defining property of life* (Roederer 2004). In this overall evolution, there is a gradual increase of complexity of the interacting systems and related mechanisms involved: from "simple" chemical reactions between biomolecules, to life-sustaining circulation systems in plants, to the ultra-complex interaction chains in the human brain.

Any information-driven interaction between *inanimate* complex systems must ultimately be life-generated or -designed, requiring at some stage goal-directed actions by a living system. Examples are the physical effects that a beaver dam has on water flow; mechanical effects on the environment of a tool used by a corvid; flight path control by an autopilot, etc. As a more explicit example, consider an electromagnetic or sound wave emitted by a meteorological lightning discharge, which does not represent any information-driven interaction: it is generated and propagates through physical processes in which information plays no role. But waves emitted by an electric discharge in the laboratory may be part of an overall artificial information-driven interaction mechanism if they are part of a device created by a human mind with the intention of having the discharge cause a desired change somewhere else, such as a record on a tape or a mental event in an observer. In summary, what distinguishes "real" life systems from purely complex physical systems is that the former, *in addition* to force-driven interactions, are capable of entertaining information-driven interactions in which the relevant triggering factor is patterns in space and time.

2 Information and Classical Physics

All information-driven interactions, whether purely biological or in a human-controlled scientific experiment or technological device, affect the "normal" non-biological course of physical and chemical events. Information-driven interactions all involve complex systems in the classical domain, with time-sequences that are compatible with causality, locality, special relativity, and thermodynamics. It is important to emphasize that information-driven interactions indeed do function on the basis of force-driven interactions between their components – what makes them different is how these purely physical components *are put together* in the interaction mechanism (the "informational architecture" of Walker, Kim, and Davies 2016).

The concept of "information" does not appear as an active, controlling agent in purely physical interaction processes in the Universe; it only appears there when a life system in general, or an *observer* in particular, intervenes (see Roederer 2005: chapter 5). In other words, "the world out there" works without information processing – until a living system intervenes and changes the physical course. The aforementioned beaver dam is an example. In physics, when we state that "a system of mass points follows a path of least action," we do not mean that the system "possesses the necessary information to choose a path of minimum action" from among infinite possibilities, but that it is *we* humans who have discovered *how* systems of mass points evolve and who developed a mathematical method applicable to all to predict or retrodict their motions. Similar arguments can be made when we describe black holes as "swallowing information," or decoherence as "carrying away information on a quantum system." In summary, the physical universe does not "obey" physical laws – it is *we the observers* who "make" the laws based on systematic observations (ultimately, specific changes in our brains – see Section 3) of the changes left in our instruments by their exposure to environmental events. And it is us who are able to devise a mathematical framework that entices us to *imagine* (see Section 3) processes like "black holes being an information-processing and -erasing system," or "an electron being at two different places at the same time."

Further examples are found in the association between entropy and information, which arises from the particular way *we* scientists describe, analyze, and manipulate nature, for instance, by counting molecules in a pre-parceled phase space; coarse-graining (averaging over preconceived domains); looking for regularities vs. disorder; quantifying fluctuations; extracting mechanical work based on observed patterns in the system; or mentally tagging molecules according to their initial states. In this latter context, let us discuss a concrete example, which also will be helpful later in a discussion of information in the quantum domain. We turn to Gibbs' paradox (for details, see Roederer 2005: section 5.5). Consider two vessels A and B of equal volume, joined by a tube with a closed valve, thermally isolated from the rest and filled with the same gas at the same pressure and temperature. If we open the valve, nothing will happen thermodynamically, but at the microscopic level, we can picture in our mind the molecules of A expanding into vessel

B and the molecules of B expanding into A. Each process would represent an adiabatic expansion with an increase of the entropy by $-kN\ln2$ (where k is Boltzmann's constant and N is the number of molecules in A or B), so there should be a total increase of the entropy of the system by $\Delta S = -2kN\ln2$ (notice the link with the entropy increase/decrease ε per bit of Shannon information: $\varepsilon = \Delta S/(2N) = -k\ln2$; see Roederer 2005). This of course is absurd and represents Gibbs paradox. In most textbooks, it is (somewhat lamely) explained away by saying that the formulas used here are indeed correct, but apply *only* if the gases in the two vessels are *physically distinguishable* (different gases, or just specimens of opposite chirality).

So, what is wrong with the conclusions of this thought experiment? By invoking the concept of pragmatic information when we say that "we can picture in our mind the molecules from vessel A doing this or that ... " we are *labeling* them so as to make them different from the others, even if in reality they are not. In other words, we are assuming that on each molecule there is a pre-established pattern (which triggers in our brain the neural correlates "this molecule is from A, that molecule is from B"). However, this pattern and any extractable information from it *do not exist* in reality – we just have forced them into our mental image! The same argumentation, invoking the concept of pragmatic information where there is none in reality, can be made regarding Maxwell's Demon paradox. Notice carefully that whenever we invoke pragmatic information in our mind, as in a *Gedankenexperiment* (i.e., establishing some imagined correlation between a pattern and a change governed by some imagined interaction mechanism), we are dealing with a thermodynamically *open system*, even if in reality it is not.

Notice that, in all these examples, the interactions involved are force-driven; but whenever we use the term "information" in their description, we really mean "pragmatic information *for us the observers*." And when *it is we the agents* who deliberately set the initial conditions of a classical mechanical system (or prepare a quantum system), we are converting it into an information-driven system with a given purpose (to achieve a change that would not happen naturally without our intervention). All laboratory experiments, whether a simple classroom demonstration or a sophisticated table-top quantum experiment, fall into this category.

It will be useful for our discussion of quantum systems to identify common features in a classical measurement process. First of all, note that it necessarily involves force-driven interactions, but they are controlled by a human being or a preplanned artifact. For instance, when you measure the size of an object with a caliper, the instrument interacts elastically with the object; if you measure it with a ruler, you need to submerge everything in a "bath of photons" whose scattering or reflection is what your optical system uses to extract the wanted information on "object–apparatus" interaction. Note that in the latter example, *the environment* has become an integral part of the instrument (this will be important for the discussion of quantum measurements). In summary, the measurement process represents an information-driven interaction between the object or system to be measured and a measurement apparatus or device (which contains a reference which we call "the unit"). The whole process *defines* the magnitude (the observable in quantum mechanics) that is being measured. And it has an observer-related purpose: that of changing his/her state of knowledge in a very specific way. It is clear then, that even in classical physics, it is impossible to detach the measurement process from the observer (or his/her measurement artifacts). In summary, the fundamental interaction stages in any measurement process are: [patterns from an object] $\rightarrow$ [change in the apparatus], and [pattern of change in the apparatus] $\rightarrow$ [change in the state of the brain of the observer], who thus has *extracted pragmatic information from the object*. Of course, in these cases that information can indeed be represented as a number (value of an observable) and all measurement procedures can be linked to Shannon or algorithmic information.

And here we should turn to another fundamental, inextricable link of the concept of information to biology, by pointing out that all the operations mentioned previously are ultimately related to how the brain of the observer reacts to external sensory input and creates internal mental images, and to how the brain of an agent plans future events and makes decisions on how to implement them. As already mentioned, these operations ultimately have the purpose of changing the neural cognitive state from "not-knowing" to "knowing," a transition that I venture to describe as the reduction of an initial brain state involving multiple expectancies to one of possible "basis states," where each basis state represents the mental image of only one possible outcome of the expected alternatives (we might even call them "preferred mental states"). Note that as a corollary, the concept of probability, usually defined mathematically as the result (limit) of a specific physical operation (e.g., a series of measurements under equal conditions like tossing dice, playing roulette), has a very subjective foundation in human brain operation.

3 Information and Brain Function

Since the beginning of quantum mechanics, physicists have been arguing about whether the observer and his/her state of knowledge, even consciousness, play an active role in the quantum measurement process. However, they did not have the benefit of knowing what is known today about the neurobiological mechanisms that control human brain function.

Recent studies with functional magnetic resonance imaging, positron emission tomography, diffusion tensor imaging, and, at the neural network level, multi-microelectrode recordings, are confirming a hypothesis long in use by neurophysicists and computer scientists, namely that the information being pro-cessed in the brain is encoded in real-time as a task-specific, *spatiotemporal distribution of neural activity* in certain regions of the cerebral cortex (e.g., Tononi and Koch 2008). If, because of neural interconnectivity, a certain specific pattern of neural activity distribution in one area triggers a specific distribution in another, and does so in a univocal way (within limits), we are in the presence of an information-based interaction between two cerebral regions; the pragmatic infor-mation involved *represents* that specific relationship, and we usually say that information has been transferred from one cerebral region to another.

When a scientist makes a measurement, the pragmatic information involved represents the correlation between an external pattern (e.g., the location in a reference space of a rigid body, the position of the dial in an instrument, the dots on cast dice, the color change of a solution) with a specific spatiotemporal pattern of neural activity in the prefrontal lobes, corresponding to the knowledge "it is *this* particular state and not any other possible one." The actual information-processing mechanisms in the brain linking one neural distribution with another are controlled by the actual synaptic wiring, which in certain regions, especially the hippocampus, has the ability of undergoing specific changes as a function of use ("plasticity") – the physiological expression of stably stored pragmatic information or long-term memory.

Modern neurobiology has an answer to the common question: when does a specific distribution of neural firings actually become a mental image? This neural activity distribution does not *become* anything – it *is* the image! (Just as the idea of information being a "purely" biological concept is unpalatable to many physicists, the idea of a mental image, even consciousness itself, being "nothing but" a very specific, unique spatiotemporal distribution of neural activity is unpalatable to many psychologists, philosophers, and theologians.) In summary, the dynamic spatiotem-poral distribution of neural impulses and the quasi-static spatial distribution of synapses and their efficiencies together are the physical realization of the global state of a functioning brain at any instant of time. Another way of expressing this: pragmatic information is encoded in the brain dynamically in short-term patterns of neural impulses and statically in the long-term patterns of synaptic architecture (it is important to point out that whenever we refer to "neural impulses" or "neural activity," we mean *any kind* of transmembrane electrical signals in neurons, whether digital – axon spikes – or analog – dendritic or cell-body activity). Given the total number of interacting elements ($\sim 10^{12}$ neurons and $\sim 10^{14} - 10^{15}$ synapses in the human brain), the number of neurons involved in one single elementary processing task (from a few to hundreds of thousands), and the discontinuous nature of activity distribution, there is little hope that a quantitative mathematical theory of integral brain function could be developed in the foreseeable future.

Quite generally, animal brains handle pragmatic information in sequences of information-driven interactions in which one specific spatiotemporal pattern of neural activity is mapped or transformed into another neural pattern – in its most basic form, from a physically triggered sensory or interoceptive stimulation pattern to a neural output pattern controlling muscle and gland fibers, thus governing the animal's integral behavior. These processes may change the interconnectivity (synaptic architecture) of participating neural networks (the learning process), leading to long-term storage of information. A memory recall consists of the replay of the original neural activity distribution that had led to the synaptic changes during memory storage; the most important type is the *associative recall*, in which the replay is triggered by a cue embedded in the ongoing neural activity distribution (for examples, see Roederer 2005). Expressed in terms of pragmatic information: in the act of remembering or imagining a certain object, information on that object stored in the synaptic architecture of the brain is retro-transferred in the form of neural impulse patterns, mostly via subcortical networks, to the visual cortex and/ or other pertinent sensory areas, where it triggers neural activity specific to the *actual* sensory perception of the object. Neuroscientists call the specific microscopic distribution of neural activity responsible for any subjective experience "neural correlate."

Let us point out at this stage that, in the definition of pragmatic information, it is often the case that different source patterns can lead to the same change in the recipient (e.g., different shapes, sizes, and colors of an apple still trigger somewhere in the cortex the neural pattern that defines the concept or image of "apple"). Likewise, the same source pattern can lead to different effects, depending on collateral information-processing activity of the system.

As a fundamentally distinct capability, the *human brain* can recall stored information *at will* as images or representations, manipulate them, discover overlooked correlations, and restore modified or amended versions thereof, and do this *without any concurrent external or somatic input* – it can go "off line" (Bickerton 1995). This is *information generation* par excellence – the establishment of univocal correspondences between mental representations (neural patterns) that have *not* been experienced through previous or real-time sensory input. It represents the *human thinking process* (e.g., Roederer 1978, 2005). Internally triggered human brain images, however abstract or "wild," are still snippets (always expressed as many different but unique patterns of neural activity in specific regions of the cortex) derived from stored information acquired in earlier sensory or mental events, and pieced together in different ways under some central control (the "main program") linked to human self-consciousness.

In animals, the time interval within which causal correlations can be established (trace conditioning) is of the order of tens of seconds and decreases rapidly if other stimuli are present (e.g., Han et al. 2003); in humans, it extends over the long-term past and the long-term future (for a brief review of human vs. subhuman intelligence, see, e.g., Balter 2010). Most importantly for our discussion, this leads to the conscious awareness of the past, present, and future, and the quantitative conception of time.

Whenever a physicist conceives or thinks about the model of a physical system or physical process, whether a mass point, a rigid body, ideal fluid, a field, a phase space or a galaxy, whether relativistic or quantum, whether one-dimensional or multidimensional, his/her brain triggers, transforms, and mutually correlates very specific and unique distributions of neural impulses. The fact that the brain is an eminently classical information-processing device (quantum decoherence times in the brain cells would be 10 or more orders of magnitude shorter than the minimum time required for any cognitive operation; see, e.g., Schlosshauer 2008) that evolved, and is continuously being trained through information-driven interactions with the classical macroscopic world, is very germane to how we can create and imagine approximate models and understand the behavior of any real (perceived or perceivable) systems, whether quantum, classical, molecular, or biological. This even applies to mathematics, from theory of sets, number theory to abstract spaces and probability theory (see Roederer 2005: section 1.6). In the latter, for instance, given a set of mental images of possible outcomes of a future quantum measurement, all may be viewed subjectively as equiprobable by an unbiased observer. Only after personal experience with multiple measurements under identical conditions, or through information from others who already have undertaken this task, can the observer develop an objective sense of traditional probability.

4 Information and Quantum Mechanics: "Bits" from "Its"

In the preceding sections we have identified two categories of interactions between bodies or systems in the universe as we know it: force-driven and information-driven. The first category is assumed to be operating in the entire spatiotemporal domain, from the Planck scale up, between elementary particles, atoms, molecules, complex condensed bodies, networks, and systems of bodies. The second category leads to the definition of pragmatic information as representing a physical, causal, and univocal correspondence between a pattern and a specific macroscopic change mediated by some complex interaction mechanism. By this very definition, the domain of validity of information-driven interactions, and therefore of the concept of pragmatic information per se, is limited to the classical macroscopic domain. The reason is that, as we shall discuss extensively later, in the subatomic quantum domain spatial or temporal patterns cannot always be defined or identified in a univocal, stable, or causal way. This in turn is related to fundamental and exclusive properties of quantum systems, which we shall examine now taking into account the definition of pragmatic information given in Section 1.

4.1 Revisiting some Relevant Quantum Facts

Specifically for the purpose of exploring the role of pragmatic information in quantum mechanics it is opportune to present, in greatly simplified fashion, some experiment-based "quantum facts in a nutshell" (an excellent and rigorous discussion of the principles of quantum mechanics relevant to this article is given in chapter 2 of Schlosshauer 2008). To emphasize that the peculiar behavior of a quantum system is not the consequence of some mathematical properties of linear algebra, Bayesian statistics, and differential equations but, rather, representative of *physical* facts happening "out there," we shall deliberately refrain from referring to abstract postulates, theorems, and properties of Hilbert spaces with which most textbooks and theories of foundations of quantum mechanics introduce the subject.

Given one *single* quantum system that has been physically prepared in a certain way, it is *impossible in principle* to determine or verify its particular state by a measurement. Indeed, having many similar quantum systems prepared in identical ways and subjecting each one to identical measurements, one will obtain a collection of measurement values from among a common (often discrete) set that only depends on the setup and the instrument. These values are called "eigenvalues" of the observable in question (as mentioned in the previous section, the observable itself being defined by the instrument). The distribution of *probabilities of occurrence* of eigenvalues, including the latter's mean value and standard deviation, is what characterizes the common state of each quantum system of the set at the time of measurement (Born's Rule). While it is impossible to predict the outcome of the measurement of any *one* system, the set of probabilities of obtaining the eigenvalues is deterministic in the sense that it depends only on the preparation and history of the quantum system and on the type of apparatus used. This is not all: when a single quantum system is measured and one of the possible eigenvalues of the corresponding observable is obtained, that quantum system emerges in a new, special state such that a repeated measurement of exactly the same kind immediately thereafter will always yield the same eigenvalue – zero standard deviation – (we exclude the so-called destructive measurements, in which the quantum system disappears, e.g., is absorbed, and we neglect any time evolution of the system). Such a state is therefore called "eigenstate" (also "basis state"). One says that the initial state of the quantum system has *collapsed* into, or been *reduced* to, an eigenstate as the result of the first intervention. In other words, after the measurement the quantum system behaves "classically" *with respect to the variable in question.*

Sometimes, the same apparatus can be used in different configurations (e.g., rotation of a Stern-Gerlach set-up or a polarization filter); experiments show that each configuration may have a different set of basis states. In other cases, there are no alternatives for the same instrumental setup (e.g., the "which-path" observable in a Mach-Zehnder interferometer; the wavelength of a spectral emission; the decay products of an unstable particle); the eigenstates are then called "preferred basis states." On the other hand, one single quantum system may have more than one observable to measure (several degrees of freedom like the position, momentum, and spin vectors of a particle); in that case, the measurement results of one observable may or may not be correlated with the distribution of measurement results of another. This means that having many multi-degree of freedom quantum systems prepared in identical ways, making the measurement of one specific observable on some and measurement of another observable on the rest, the statistical distributions of results in both series of measurements may or may not be correlated (non-commuting or commuting variables, respectively).

It is when non-commuting observables enter the picture that Planck's universal constant $\hbar$ appears and delimits the "size" of the quantum domain. Historically, it was the position and the momentum of a particles that led Heisenberg to the uncertainty relationship between their standard deviations Δx, Δp in measurements under similar conditions ($\Delta x \Delta p \geq \hbar/2$). For the sake of completeness of this summary, we must also mention that Schrödinger's equation is introduced when the time evolution of a quantum system is to be described (which, however, will not be dealt with here).

All these experimental facts have allowed the development of a comprehensive picture of linear algebra algorithms in which general states of a quantum system are represented by (unit) vectors in a Hilbert space; interactions such as measurements and transformations (e.g., mutual interactions, time evolution) are represented by specific operators. Each observable defines a subspace whose axes represent its eigenstates or basis states when measured with the observable-defining apparatus in a specific configuration, and the squares of the projections of the state vector onto the axes represent the probabilities of occurrence of the corresponding eigenvalues. A general quantum state (called "pure" or "superposed" state) is thus represented as a *linear superposition of basis states*; a measurement is represented by the rotation of the state vector onto one of the axes. Complex numbers are used for the state vector components to accommodate the possibility of having two or more superposed states with the same probability distribution of measurement results, but that behave differently in interactions (interference phenomena). Finally, the functional expressions (e.g., the Hamiltonians) governing the dynamic evolution of the observables of a quantum system are equivalent, in many cases identical, to the corresponding functional expressions in a classical system.

Here comes our first real encounter with the concept of pragmatic information. The mathematical formalism developed for quantum systems in correspondence with the formalism of classical mechanics tempts us to picture *in our mind* a superposed state as the quantum system "being in different eigenstates at the same time." As stated before, for a *single* quantum system it is impossible to determine through measurement whether it is in such a superposition, and if according to its history it should be, what that superposition really is. In other words: it is impossible to extract pragmatic information from a single quantum system. This is no different than wanting to imagine in the aforementioned Gibbs paradox the molecules from one container moving into the other preserving a distinguishability as if they were carrying labels from which to extract pragmatic information – when in reality there is none.

Let us consider an ensemble of N equally prepared, identical quantum systems. We measure the same variable on each one and compile the results, finding a whole set of different outcomes in a certain statistical distribution. This tells us that before being measured, the *preparation* process had produced an ensemble of quantum systems, each one in *the same coherent superposed state* characterized by that statistical distribution. *After* the measurements are made, we are left with a radically different kind of ensemble, namely a *mixture* of N systems in which each component is now collapsed into a generally different, but repetitively measurable eigenstate (remember that in all this we are ignoring possible changes as a function of time). In each case, the component can be *labeled* with the pertinent result (this process is called "filtering"); this tells us that the mixture would behave classically with respect to the observable in question, because now we *can* extract pragmatic information from it (e.g., we could create stable patterns and establish univocal correspondences with changes or new patterns elsewhere). Notice again that the big difference with classical physics is that after the preparation but before the measurements, the elements of the ensemble are each in the *same* state – this is not the case if we were to run ("prepare"), say, N pinball machines or roulettes! We already mentioned that if we let the equally prepared elements interact in the quantum domain *without* performing any measurements (interactions with macroscopic systems), experiments show that all elements indeed emerge from preparation in identical (but superposed) states.

Next, we consider an ensemble of several *different* non-interacting quantum systems in different superposed states (i.e., with different histories). Each one is described by its own separate state vector in its own Hilbert space (corresponding to the observable(s) chosen). If measured individually, we would obtain individual results for each component uncorrelated from each other. Let us now bring these individual quantum systems into mutual interaction at the quantum level (i.e., shielded from interactions with macroscopic bodies), take them away from each other, and only then make the measurements. Like before, we cannot predict what the individual outcomes will be, but once obtained, experiment tells us that *they appear correlated* – regardless where they were taken after the interaction, and in which temporal order they were measured. This means that once a set of quantum systems in superposed states interact, they lose their independence: the ensemble behaves as *a single whole*, non-local in space and time, described by just *one* global composite state. The component parts are then said to be *entangled*. The actual results of measurements on entangled components (unpredictable in themselves, except statistically) will be correlated, no matter *how far away in space and time* they are located from each other after the mutual interaction (there is no equivalent example of this in classical physics). It is often stated: "once the measurement has been made on one system, the other one will automatically collapse into a conjugate basis state." Yet there is no actual information transfer involved in this process, no super-luminous communication between the two – there is no way to deliberately craft a pattern with some components and then expect that a correlated pattern appears in the rest! All we find (at the macroscopic level) is that the measurement results will be correlated – but correlation does not imply causation!

It is also an experimental fact that not only man-made laboratory measurements but in effect *any* interaction with the natural macroscopic environment will eventually break up the global state of an ensemble of entangled quantum systems into independent, ultimately collapsed states of its individual components. This process is called "decoherence" (e.g., Schlosshauer 2008). As a matter of fact, it is impossible to completely shield a quantum system from unpredictable macroscopic influences of the environment, and superposed quantum states are difficult to maintain in the laboratory.

Entanglement allows us to take a better look at the quantum measurement process, in which a quantum system is deliberately made to interact with a macroscopic device. This device is constructed in such a way that, in the special case when the quantum system is in an eigenstate of the observable that the apparatus is supposed to measure, the initial perturbation propagates through the instrument and is amplified to give an observable macroscopic effect (e.g., the particular position of a dial) that depends on the initial eigenstate. When the quantum system to be measured is now in a superposed state, it gets entangled with the "quantum end" of the apparatus – but as subsequent interactions propagate through the instrument, inevitable entanglement with the environment will eventually cause the single quantum state of the total system (measured system plus instrument) to break up into mutually correlated basis states: (i) the instrument signals a specific macroscopic change, and (ii) the original quantum system emerges in a corresponding eigenstate (the "Schrödinger Cat" paradox refers to what would happen if there was no decoherence in this process). In this whole process, the only information-driven interaction occurs between the "classical end" of the instrument and the observer (sensory effect of the macroscopic change in the instrument on the observer's brain, in real time or after being exposed to a human-designed recording device).

When an ensemble of quantum systems has decohered, it will behave classically, at least with respect to the observables whose states have collapsed, furnishing the *same* results in immediately successive, identical measurements. Only then can we define patterns based on measurement outcomes for information-driven interactions – such interactions cannot occur at the non-decohered pure-state quantum level. We can emphasize again: *pragmatic information cannot exist and operate in, or be extracted from, a pure quantum domain.* A physical consequence is that a quantum system in a superposed state cannot be copied (the so-called no-cloning theorem). We'll come to this again in the next section.

It is an experimental fact that macroscopic bodies are in decohered and reduced states, particularly of the position variable – down to the domain of atomic nuclei of organic molecules; but not their subatomic constituents such as electrons, nucleons, quarks (this does not mean that under very special laboratory conditions entire molecules cannot find themselves as a whole in a superposed quantum state – diffraction experiments with C^{60} molecules have been carried out successfully; see Hackermüller et al. 2004). Molecules behave classically and can carry, transfer, or respond to pragmatic information: a spatial pattern in some organic molecule (e.g., RNA) triggering a specific change in some other molecular system (e.g., formation of a protein) is possible! What is required as a condition *sine qua non* for this to happen is the intervention of some complex molecular organelle (e.g., a ribosome) responsible for the interaction mechanism. Herein lies the essence of the evolutionary emergence of life systems (Roederer 2003, 2005, 2016).

Let us now walk through some well-known examples, usually taught at the beginning of a course on quantum computing, and examine them strictly from the point of view of pragmatic information – pointing out where the latter does play a role and where it can't.

4.2 Single Qubits

Consider the measurement of a qubit (elementary quantum system with only two possible eigenstates, like a half-spin silver atom, a photon in a two-path Mach-Zehnder interferometer). The qubit interacts with the apparatus at its quantum end

128 *Juan Roederer*

(e.g., the atoms to be excited or ionized in a particle detector), and after the measurement the classical end exhibits a macroscopic change (e.g., voltage pulse, a blip on a luminescent screen, position of a pointer, a living or dead cat). It is the physical structure of the device enabling the occurrence of such macroscopic change that *defines* the observable in question (although in this case it is better to say that, given the qubit, its two preferred basis states determine the instrument to be used). The observer per se is irrelevant during the measurement itself, except that we must not forget that it was a human being who set up the measurement (hence decided what observable to measure), who selected and prepared the qubit to be measured, and whose brain ultimately expects to receive an image, the neural correlate, of the change of the macroscopic state of the apparatus as a result of the measurement (knowledge of the value of the binary observable).

Following the usual von Neumann protocol, let us call $|M\rangle$ the initial state of the apparatus and $|M_0\rangle$, $|M_1\rangle$ the two possible alternative states of the apparatus *after* the measurement. The instrument is *deliberately* built in such a way that when the qubit to be measured is in basis state $|0\rangle$ before its interaction with the apparatus, the final independent state of the instrument after the measurement will be $|M_0\rangle$, and if the state of the qubit is $|1\rangle$, the instrument will end up in state $|M_1\rangle$. In either case, the state of the qubit remains unchanged (again, we are assuming this to be a non-destructive measurement). Therefore, for the *composite* state qubit-apparatus we will have the following evolution in time: $|0\rangle|M\rangle \rightarrow |0\rangle|M_0\rangle$ or $|1\rangle|M\rangle \rightarrow |1\rangle|M_1\rangle$, respectively.

If the qubit is now in a *superposed state* $\alpha|0\rangle + \beta|1\rangle$ (with $|\alpha|^2 + |\beta|^2 = 1$), taking into account what we said in Subsection 4.1, we will obtain an *entangled* state and the state of the composite system qubit-apparatus will remain a linear superposition, as long as it is kept isolated from all other interactions: $(\alpha|0\rangle + \beta|1\rangle)|M\rangle \rightarrow \alpha|0\rangle|M_0\rangle + \beta|1\rangle|M_1\rangle$. However, it is an experimental fact that one has never observed a macroscopic system with such peculiar properties as superposition (the essence of the Schrödinger Cat paradox): the end state of the composite system will always be either $|0\rangle|M_0\rangle$ or $|1\rangle|M_1\rangle$ (decoherence) with $|\alpha|^2$ and $|\beta|^2$ the probabilities to obtain either result, respectively, in a large set of measurements under strictly identical conditions of preparation (Born Rule). In each process, the instrument comes out in the macroscopic state $|M_0\rangle$ or $|M_1\rangle$, and the original qubit emerges reduced to the corresponding eigenstate $|0\rangle$ or $|1\rangle$.

According to our definition of pragmatic information, decoherence and state reduction thus express the fundamental fact that *no information can be extracted experimentally on the superposed state of a single qubit*. A direct consequence, already mentioned before, is that the superposed state of one given qubit cannot be copied. Let us restate what we said before: if we *could* make $N(\rightarrow \infty)$ copies of a single qubit in a superposed state, a correspondence could indeed be established between the original pair α, β and some macroscopic feature linked to the statistical outcome of measurements on the N copies (in the case of a qubit, e.g., the average values of some appropriate observables). This would be tantamount to extracting pragmatic information from the original *single* qubit. What is possible, though, is to repeat exactly N times the *preparation process* to obtain N separate qubits in the same superposed state (like retyping a text on blank sheets repeatedly, instead of Xeroxing the original page N times). Each qubit of this set can then be subjected to a measurement, and the parameters α, β extracted from the collection of results; this is, precisely, how the probabilities $|\alpha|^2$ and $|\beta|^2$ are obtained experimentally (remember that, since α and β are two normalized *complex* numbers, in order to determine, say, their relative amplitudes and phase statistically, it is necessary to obtain *two* sets of measurement data $-N/2$ measurements in each set).

Usually, the *preparation* of a qubit in a given superposed state requires the intervention of complex macroscopic devices and three steps of action on a *set* of qubits: 1) measurement of an appropriate observable (which leaves each qubit in an eigenstate of that observable); 2) selection (filtering) of the qubits that are in the desired eigenstate (a classical process); 3) unitary transformation (a rotation in Hilbert space) to place the selected qubit in the desired superposed state in other bases. In this procedure, a preparer has converted a certain *macroscopic* pattern (embedded in the physical configurations of the preparation process) into the values of two complex parameters of a quantum system (e.g., the normalized α and β coefficients in the mathematical description of the qubit). According to our definition, doesn't such correspondence represent genuine pragmatic information (correspondence pattern $\rightarrow$ change)? No, because it would not be univocal: given a *single* qubit, an agent could never reconstruct through measurement the original macroscopic configuration used, or the steps taken, in the preparation process. The only way to do so is to remain in the macroscopic domain and *ask* the preparer (classical information from brain to brain!).

4.3 Entangled Qubits and Space-Time

Take two qubits A and B that through some preparation or creation process are maximally entangled in the antisymmetric Bell state $\Psi^- = 1/\sqrt{2}(|0\rangle_A|1\rangle_B - |1\rangle_A|0\rangle_B)$ at time t_0. We may imagine qubit B now being taken far away. If nothing else is done to either, we can bring B back, and with some suitable experiment (e.g., interference) demonstrate that the total state of the system had remained entangled all the time. If, instead, at time $t_A > t_0$ a measurement is made on qubit A leaving the composite system reduced to either state $|0\rangle_A|1\rangle_B$ *or* $|1\rangle_A|0\rangle_B$, qubit B will appear in either basis state $|1\rangle_B$ *or* $|0\rangle_B$, respectively, if measured. The puzzling thing is that it does not matter *when* that measurement on B is made – even if made *before* t_A. Of course, we cannot predict which of the two alternatives will result; all we can affirm is that the measurement results on each qubit will appear to be correlated, no matter the mutual spatial distance and the temporal order in which these were made. One could argue that in the case of a measurement on B at an earlier time $t_B < t_A$, it was *this* measurement that "caused" the reduction of the qubits' composite state – but the concepts of "earlier" and "later" between distant events are not relativistically invariant properties (see also next paragraph).

The result of all this is that it appears as if the reduction of the quantum state of an entangled system triggered by the measurement of one of its components is "non-local in space and time." Yet as stated before, correlation does not mean causation in the quantum domain: nothing strange happens at the macroscopic level: the state reduction cannot be used to transmit any real information from A to B. In terms of our definition of pragmatic information, there is no "spooky" action-at-a-distance: an agent manipulating A has no control whatsoever over *which macroscopic change* shall occur in the apparatus at B, and vice versa. The spookiness only appears when, in the *mental* image of a pair of spatially separated entangled qubits, we force our (macroscopic) concept of information into the quantum domain of the composite system and think of the act of measurement of one of the qubits as *causing* the particular outcome of the measurement on the other.

Yet another insight can be gleaned from the reexamination of the so-called *quantum teleportation* of a qubit (e.g., Bouwmeester et al. 1997). Let us consider the basic experimental protocol: an entangled pair of qubits in the antisymmetric Bell state $\Psi^- = 1/\sqrt{2}(|0\rangle_A|1\rangle_B - |1\rangle_A|0\rangle_B)$ is produced at time t_0 and its components are taken far away from each other. At time $t_A > t_0$ an unknown qubit in superposed state $\alpha|0\rangle_C + \beta|1\rangle_C$ is brought in and put in interaction with qubit A. The total, composite state of the three-qubit system is now $\Psi^- = 1/\sqrt{2}(\alpha|0\rangle_C + \beta|1\rangle_C)(|0\rangle_A|1\rangle_B - |1\rangle_A|0\rangle_B)$, which can be shown algebraically to be equal to a linear superposition of four Bell states in the A-C subspace, with coefficients that are specific unitary transforms of the type $(-\alpha|0\rangle_B - \beta|1\rangle_B)$, $(-\alpha|0\rangle_B + \beta|1\rangle_B)$, and so on. Therefore, if a measurement is made on the pair A-C of any observable whose eigenstates are the four Bell states, the state of the entire system will collapse into just one of the four terms, with the qubit at B left in a superposed state with coefficients given by the parameters α, β of the now vanished unknown qubit C. If the observer at point A informs B (a classical, macroscopic transfer of information) which basis Bell state has resulted in the measurement – only two bits are needed to label each possible basis state – observer B can apply the appropriate inverse unitary transformation to his qubit, and thus be in possession of the teleported qubit C (defined by the unknown coefficients α and β).

The puzzling aspect of this procedure is that it looks as if the infinite amount of information on two real numbers (those defining the normalized pair of complex numbers α and β) was transported from A to B by means of only two classical bits. The answer is that, again, according to our definition, α and β *do not represent pragmatic information* on the state of any *given* qubit. They are quantitative parameters in the mathematical framework developed to describe quantum systems and their interactions, but they cannot be determined physically "out there" for a given qubit. Related to this, there is no way to verify the teleportation of a *single* qubit; the only way verification could be accomplished is through a *statistical* process, repeating the whole procedure N times, from the identical preparation of each one of the three qubits A, B, C to the actual measurement of the teleported qubit. If we determine the frequencies of occurrence N_0 and $(= N - N_0)$ of the $|0\rangle_B$ and $|1\rangle_B$ states, and express the teleported qubit state in its polar (Bloch sphere) form $|\Psi\rangle = cos(\theta/2)|0\rangle + exp(i\varphi)sen(\theta/2)|1\rangle$, it can be shown that the number of *statistically significant* figures (in base 2) of θ and φ to be obtained is equal to the *total number* $2N$ of bits transmitted classically (i.e., macroscopically) from A to B. Therefore, from the *statistical* point of view, there is no puzzle at all! Moreover, this analysis shows that there is *no way* of teleporting pragmatic information and, as a consequence, macroscopic objects!

The preceding discussion also reveals something about how we tend to think intuitively of time and space at the quantum level. A point in the four-dimensional continuum of space-time is a mathematical abstraction, useful in the description of objects and events in the Universe. But *position* and *time intervals* of objects must be determined by measurements, i.e., with a macroscopic instrument, in which a macroscopic change is registered (e.g., a change in the "bath of light" in a position measurement; a change in the geometric configuration of a clock (its hands) or the angular position of a star). Like information, *time is a macroscopic concept* (even an atomic clock must have classical components to serve as a timepiece). We can assign time marks to a quantum system only when it interacts *locally* with (or is prepared by) a macroscopic system. In the case of a wave function $\Psi(x, t)$, the time variable refers to the time, measured by a macroscopic clock *external* to the quantum system, at which $|\Psi|^2$ is the probability density of actually observing a quantum system at the position x in configuration space, which is also based on a measurement with a macroscopic instrument.

Non-locality in space and time really means that for a composite quantum system the concepts of distance and time interval between different superposed or entangled components *are undefined* as long as they remain unobserved, i.e., free of interactions with macroscopic systems. Already in 1927 Heisenberg declared: "A particle trajectory is created only by the act of observing it"! (for a postulate on atemporal evolution, see Steane 2007). Because of this, it may be unproductive trying to find a modified form of the Schrödinger equation, or any other formalism, to describe quantitatively what happens "inside" a quantum system during the process of state reduction (e.g., a theoretical or experimental derivation of "average trajectories" in a double-slit experiment – e.g., Kocsis et al. 2011 – provides the geometric visualization of something on which, for an *indivi-dual* particle, pragmatic information could never be obtained!).

But, finally, what about the decay process of an unstable particle (or nucleus, for that matter), which runs on the proper time of the particle, as demonstrated long ago with the decay times of cosmic ray *mu*-mesons when observed from a reference frame fixed to Earth? As we shall briefly mention, decay processes may be linked to decoherence (reduction to decay products); if that is indeed the case, they would be influenced by interactions with the macroscopic environment (the Universe) *as experienced by the quantum system*, not the observer.

4.4 The Process of Decoherence

Let me briefly address the still contentious question of what, if any, *physical* processes are responsible for the transition from the quantum end of a measurement apparatus to its classical, observable one (for details, see, e.g., Schlosshauer 2008). For this purpose, let us consider a super-simplified *model* of a measuring apparatus that consists entirely of mutually interacting qubits – lots of them, perhaps 10^{22} to 10^{25} – with one of which the external qubit to be measured enters into unitary non-destructive interaction at time t_0. In our model, the apparatus qubits represent a complex web in some initial or "ready" state, designed in such a way that, as the local unitary interaction processes multiply and propagate, only two distinguishable macroscopic end states M_0 and M_1 can be attained, realized as two macroscopic spatially or temporally different forms or patterns (the so-called pointer states, represented in mutually orthogonal Hilbert subspaces of enormous dimensions). The key physical property of this construction is that, for a qubit in a basis state, the instrument's final macroscopic configuration will depend on the *actual* basis state of the measured qubit (see discussion in Subsection 4.2).

If the qubit to be measured is now in a superposed state, the first physical interaction at time t_0 would create an entangled state of the system "qubit-first apparatus quantum element," which through further unitary inter-component inter-actions would then expand to the entire composite system "qubit-apparatus" in a cascade of interactions and further entanglements throughout the apparatus. Accordingly, the classical end of the apparatus also should end up in a superposed state (the Schrödinger Cat!) and, since in principle the interactions are unitary, the whole process would be reversible. Obviously, somewhere in the cascade of interactions there must be an irreversible breakdown of the entanglement between the original qubit and the instrument, both of which will emerge from the process in separate but correlated states, either $|0\rangle$ and M_0, or $|1\rangle$ and M_1, respectively. Extractable pragmatic information appears only at the classical end of this cascade. All this means that if exactly the same kind of measurement is repeated immediately on the qubit (assuming that it was not destroyed in the measurement process), one will obtain a result that *with certainty* is identical to that of the original measurement.

Measurement processes and their apparatuses are *artifices* – human planned and designed for a specific purpose. However, note that the preceding discussion on decoherence can be applied to the case in which we replace the artificial measurement apparatus with the *natural environment per se*. Just replace the word "apparatus" with "the environment" with which a given quantum system willy-nilly interacts and gets entangled. As long as this entanglement persists, the given quantum system will have lost its original separate state, and only the *composite* quantum system-environment will have a defined state, however complicated and delocalized. Now, if the interaction with the environment leads to a *macroscopic* change somewhere (potentially verifiable through classical information extraction by an observer, but independently of whether such verification actually is made), it will mean that decoherence has taken place and that the state vector of the original quantum system will have been reduced to one of its original eigenstates pertinent to the particular interaction process.

This is usually described as "entanglement with the environment carrying away information on a quantum system," or "information about the system's state becoming encoded in the environment." However, I would like to caution about the use of the term "information" in this particular context: there is no loss of pragmatic information in natural decoherence, because there wasn't any there in the first place! A much less subjective way is to say: *a quantum system continuously and subtly interacts with its environment and gets entangled with it; if decoherence occurs, a macroscopic change in the state of the environment will appear somewhere (information about which could eventually be extracted by an observer), and the state of the quantum system will appear reduced to some specific basis state in correspondence with the environmental change in question.* Since this basis state is in principle knowable, the decohered qubit now belongs to the classical domain. Measurement instruments are environmental devices specifically designed to precipitate decoherence and steer them into certain sets of possible final macroscopic states (preferred states, if no alternative sets exist).

An ensemble of identically prepared quantum systems (e.g., a chunk of a recently separated, chemically pure radioisotope) thus turns probabilistic because it is unavoidably "submerged" in a gravitating, fluctuating, thermodynamic macro-world, and will decay into a mixture of quantum entities in eigenstates (e.g., with an α-particle either still inside or already outside a nucleus), the decay times mainly depending on the original wave function of the individual nuclei, but also slightly perturbed by subtle but unavoidable interactions of the latter with the environment (leading to fluctuations in the exponential decay of the ensemble). On the other hand, the laboratory measurement of a quantum system may be viewed as a case in which the environment was deliberately altered by interposing a human-made apparatus, which then altered in a "not-so-subtle" way the time evolution of the system (we should really say: the time evolution of potential macroscopic effects of the system) by greatly increasing the chance of decoherence. In summary, what we have called the cascade of entanglements in a quantum measurement also involves a stochastic ensemble of outer environmental components with which the instrument's components are in subtle but unavoidable interaction.

A collateral consequence of natural decoherence is that any peculiar quantum property like superposition will have little chance of spreading over a major part of a macroscopic object, which indeed will behave classically whenever observed – there always seems to be a natural limit to the complexity of a quantum system in a pure superposed state beyond which it will decohere. In other words, the classical macroscopic domain, in which life systems operate and information can be defined objectively, consists of objects whose constituents have decohered into eigenstates (mainly, of their Hamiltonians). Quantum behavior of a macroscopic system is *not* forbidden (a Schrödinger Cat *could* be in a superposed state of dead and alive at the same time!), but its probability and duration would be ridiculously small. This also explains the fact that, as mentioned before, many artificial quantum systems are very unstable in a superposed state, and thus very difficult to handle in the laboratory – a fact that represents one of the biggest challenges to quantum computing.

Finally, we should view all dynamics equations in quantum mechanics, like the Schrödinger equation, as the tools for providing information on potential *macroscopic* effects of a quantum system on the environment (or a measurement apparatus) under given circumstances, rather than describing the evolution of the quantum system per se. The "amazing aspect" of quantum mechanics is not its puzzling paradoxes, but the fact that a mathematical framework *could* be developed that can be used successfully to determine statistical, objective probabilities for *observable* macroscopic outcomes of interaction processes both, in artificial experiments and in the natural environment.

5 Concluding Remarks: Quantum Pedagogy

From the previous discussion, it is advisable to refrain from using the classical concept of pragmatic information indiscriminately to represent mentally a quantum system – be it by just thinking about it, mathematically representing it, or manipulating it in *Gedankenexperiments*. Yet quite commonly we do, especially when we teach – but then, as mentioned previously, we should not be surprised that by *forcing* the concept of information into the quantum domain, mental images are triggered of "weird" behavior that is contradictory to our everyday macroscopic experience.

To me, the problem of the interpretation of quantum mechanics is not just one of a philosophical nature, but an eminently pedagogical one. For instance, how should one answer correctly the often-asked question: Why is it not possible, even in principle, to extract information on the actual state of a *single* qubit? Because *by the definition of information*, to make that possible there would have to exist some physical paradigm by means of which a change is produced somewhere in the *macroscopic* classical domain that is in one-to-one correspondence with the qubit's parameters immediately prior to that process. Only for eigenstates (basis states) can this happen – decoherence prevents the formation of any macroscopic trace of superposed states. In the case of an initially superposed state, the end state of the qubit will always appear correlated with the end state of the macroscopic system, i.e., will emerge reduced to a correlated or preferred basis state. In somewhat trivial summary terms, quantum mechanics can only provide real information on natural or deliberate *macroscopic imprints* left by a given quantum system that has undergone a given preparation, eventually interacts unitarily (reversibly) with other quantum systems forming a composite quantum system, which as a single whole interacts irreversibly with the surrounding macroscopic world.

So what are the coefficients in a qubit state like $|\Psi\rangle = \alpha|0\rangle + \beta|1\rangle$ or, $|\Psi\rangle = cos(\theta/2)|0\rangle + exp(i\varphi)sen(\theta/2)|1\rangle$? They are *parameters in a model representation* of the system in complex Hilbert space, which with an appropriate mathematical framework enables us to make quantitative, albeit *only probabilistic*, predictions about the system's possible macroscopic imprints on the classical domain. We may call α and β or θ and φ "information," and we do, based on the fact that we can prepare a quantum system in a chosen superposed state – the common usage of the terms "quantum bit" and "quantum information" testifies to this. Yet for a *single* qubit we cannot retrieve, copy, or verify the numbers involved, which means we cannot establish a univocal correlation between the state of the qubit and any macroscopic feature. In other words, *those parameters are not pragmatic information* (the only exception is when the qubit is in one of its basis states). This is why when we do call α and β "information," we are always obliged to point out its "hidden nature"! And in teaching, we always would have to mumble something about super-luminal speed of information, teleportation of real things, a particle being in different positions at the same time, etc. to satisfy our (classical world) imagination. In Richard Feynman's words, we always would have to emphasize that "the [quantum] 'paradox' is only a conflict between reality and your feeling of what reality 'ought to be.'" The whole framework of quantum information theory and computing is based on the consistency of the following kind of *classical* correspondence: the relation of a given initial set of qubits in prepared eigenstates (a classical input pattern) correlated through intermediate unitary quantum interactions in an appropriately shielded quantum computing device with another final set of qubits in basis states (a classical output pattern). This input-output correlation is what really should be called "quantum information," a genuine category of pragmatic information.

Obviously, during the time interval between input and output, any extraneous non-unitary intervention, whether artificial (a measurement) or natural (decoherence), will change or destroy the macroscopic input-output correlation. Indeed, in this interim interval, the proverbial mandate of "don't ask, don't tell" applies (Roederer 2005) – not because we don't know *how* to extract relevant information to answer our questions, but because pragmatic *information per se does not operate in the quantum domain.*

Let me end by stating a personal opinion as a former physics teacher. When it comes to evaluating, or to teaching about, the "realities out there," it is the physicists who should "get real" and recognize the fact that the World, both physical and biological, does not operate on the basis of what happens in Mach-Zehnder inter-ferometers, Stern-Gerlach experiments, two-slit diffraction laboratory setups, qubit teleportation, and all these marvelous experiments designed and performed by humans. In reality, all such experiments, while providing answers to the inborn human inquiry about how our environment works, are but artificial intrusions poking into a Universe that does not care about linear algebra, Hamiltonians, and information per se. These experiments and the ensuing human understanding were made possible only because of the emergence, at least on Planet Earth, of interactions based not on force fields alone, but on the evolution of ultra-complex macroscopic mechanisms responding exclusively to simple geometric patterns in space and time.

References

Balter, M. (2010). "Did Working Memory Spark Creative Culture?" *Science*, **328**: 160–163.

Bickerton, D. (1995). *Language and Human Behavior*. Seattle: University of Washington Press.

Bouwmeester, D., Pan J. W., Mattle, K., Eibl, M., Weinfurter, H., and Zeilinger, A. (1997). "Experimental Quantum Teleportation." *Nature*, **390**: 575–579.

Hackermüller, L., Hornberger, K., Brezger, B., Zeilinger, A., and Arndt, M. (2004). "Decoherence of Matter Waves by Thermal Emission of Radiation." *Nature*, **427**: 711–714.

Han, C. J., O'Tuathaigh, C. M., Van Trigt, L., Quinn, J. J., Fanselau, M. S., Mongeau, R., Koch, C., and Anderson, D. J. (2003). "Trace but not Delay Fear Conditioning Requires Attention and the Anterior Cingulated Cortex." *Proceedings of the National Academy of Sciences, USA*, **100**: 13087–13092.

Kocsis, S., Braverman, B., Ravets, S., Stevens, M. J., Mirin, R. P., Shalm, L. K., and Steinberg, A. M. (2011). "Observing the Average Trajectories of Single Photons in a Two-Slit Interferometer." *Science*, **332**: 1170–1173.

Küppers, B. O. (1990). *Information and the Origin of Life*. Cambridge, MA: MIT Press.

Roederer, J. G. (1978). "On the Relationship Between Human Brain Functions and the Foundations of Physics." *Foundations of Physics*, **8**: 423–438.

Roederer, J. G. (2003). "Information and Its Role in Nature." *Entropy*, **5**: 1–31.

Roederer, J. G. (2004). "When and Where Did Information First Appear in the Universe?" Pp. 23–42 in *New Avenues in Bioinformatics*, J. Seckbach and E. Rubin (eds.). Dordrecht: Kluwer Academic Publishers.

Roederer, J. G. (2005). *Information and Its Role in Nature*. Berlin, Heidelberg, New York: Springer-Verlag.

Roederer, J. G. (2016). "Pragmatic Information in Biology and Physics." *Philosophical Transactions of the Royal Society A*, **374**: 20150152.

Schlosshauer, M. (2008). *Decoherence and the Quantum-to-Classical Transition*. Berlin-Heidelberg: Springer-Verlag.

Shannon, C. E. and Weaver, W. (1949). *The Mathematical Theory of Communication*. Champaign: University of Illinois Press.

Steane, A. M. (2007). "Context, Spacetime Loops and the Interpretation of Quantum Mechanics." *Journal of Physics A: Mathematical and Theoretical*, **40**: 3223–3243.

Tononi, G. and Koch, C. (2008). "The Neural Correlates of Consciousness: An Update." *Annals of the New York Academy of Science*, **1124**: 239–261.

Walker, S. I., Kim, H., and Davies, P. C. W. (2016). "The Informational Architecture of the Cell." *Philosophical Transactions of the Royal Society A*, **374**: 20150057.

Wheeler, J. A. (1989). "Information, Physics, Quantum: The Search for Links." Pp. 354–368 in H. Ezawa, S. I. Kobayashi, and Y. Murayama (eds.), *Proceedings III International Symposium on Foundations of Quantum Mechanics*. Tokyo: Physical Society of Japan.

7

Interpretations of Quantum Theory: A Map
of Madness

ADÁN CABELLO

1 Introduction

Motivated by some recent news (Hensen et al. 2015), a journalist asks a group of physicists: "What's the meaning of the violation of Bell's inequality?" One physicist answers: "It means that non-locality is an established fact." Another says: "There is no non-locality; the message is that measurement outcomes are irreducibly random." A third one replies: "It cannot be answered simply on purely physical grounds; the answer requires an act of metaphysical judgment." (All of our characters are fictitious. However, it might be interesting to compare their points of view with those in Gisin 2012, Zeilinger 2005, and Polkinghorne 2014, respectively). Puzzled by the answers, the journalist keeps asking questions about quantum theory: "What is teleported in quantum teleportation?" "How does a quantum computer really work?" Shockingly, for each of these questions, the journalist obtains a variety of answers that, in many cases, are mutually exclusive. At the end of the day, the journalist asks: "How do you plan to make progress if, after 90 years of quantum theory, you still don't know what it *means*? How can you possibly identify the *physical* principles of quantum theory or expand quantum theory into gravity if you don't agree on what quantum theory *is about*?"

Here we argue that it is becoming urgent to solve this too long-lasting problem. For that, we point out that the interpretations of quantum theory are, essentially, of two types and that these two types are so radically different that there *must be* experiments that, when analyzed outside the framework of quantum theory, lead to different empirically testable predictions. Arguably, even if these experiments do not end the discussion, they will add new elements to the list of strange properties that some interpretations must have, therefore they will indirectly support those interpretations that do not need to have all these strange properties.

2 The Many Interpretations of Quantum Theory

As Mermin points out:

[Q]uantum theory is the most useful and powerful theory physicists have ever devised. Yet today, nearly 90 years after its formulation, disagreement about the meaning of the theory is stronger than ever. New interpretations appear every day. None ever disappear.

(2012: 8)

This situation is odd and is arguably an obstacle for scientific progress, or at least for a certain kind of scientific progress. The periodic efforts of listing and comparing the increasing number of interpretations (Belinfante 1973; Jammer 1974; Bub 1997; Dickson 1998) show that there is something persistent since the formulation of quantum theory: interpretations are essentially of two types; those that view quantum probabilities of measurement outcomes as determined by intrinsic properties of the world and those that do not. Here we call them "type-I" and "type-II," respectively. In Table 7.1 some interpretations are classified according to this criterion and some extra details are given.

Table 7.1. *Some interpretations of quantum theory classified according to whether they view probabilities of measurement outcomes as determined or not by intrinsic properties of the observed system.*

	ψ-Ontic	ψ-Epistemic
Type I (intrinsic realism)	Bohmian mechanics (Bohm 1952; Goldstein 2013)	Einstein (Einstein 1936)
	Many worlds (Everett 1957; Vaidman 2015)	Ballentine (Ballentine 1970)
	Modal (Van Fraassen 1972; Lombardi and Dieks 2012)	Consistent histories (Griffiths 1984, 2014)
	Bell's "beables" (Bell 1976)	Spekkens (Spekkens 2007)
	Collapse theories* (Ghirardi, Rimini, and Weber 1986; Ghirardi 2011)	

* Collapse theories modify or supplement the unitary formalism of quantum theory; therefore, they are not pure interpretations.

	About knowledge	About belief
Type II (participatory realism)	Copenhagen (Bohr 1998; Faye 2014)	QBism (Fuchs 2010; Fuchs and Schack, 2013; Fuchs, Mermin, and Schack 2014)
	Wheeler (Wheeler 1983, 1994)	
	Relational (Kochen 1985; Rovelli 1996)	
	Zeilinger (Zeilinger 1999, 2005)	
	"No-interpretation" (Fuchs and Peres 2000)	
	Brukner (Brukner 2016)	

"Type-I interpretations" are defined as those in which the probabilities of measurement outcomes are determined by intrinsic properties of the observed system. Type-I interpretations can be "ψ-ontic" (Harrigan and Spekkens 2010) if they view the quantum state as an intrinsic property of the observed system, or "ψ-epistemic" (Harrigan and Spekkens 2010), if they view the quantum state as representing knowledge of an underlying objective reality in a sense somewhat analogous to that in which a state in classical statistical mechanics assigns a probability distribution to points in phase space. "Type-II interpretations" are defined as those that do not view the probabilities of measurement outcomes of quantum theory as determined by intrinsic properties of the observed system. Type-II interpretations do not deny the existence of an objective world, but, according to them, quantum theory does not deal directly with intrinsic properties of the observed system, but with the experiences an observer or agent has of the observed system. Type-II interpretations can be "about knowledge" if they view the quantum state as an observer's knowledge about the results of future experiments, or "about belief" if they view the quantum state as an agent's expectations about the results of future actions. This table is based on a similar table made by Leifer (2014), but has many significant differences and incorporates suggestions from many colleagues (see acknowledgments). The term "participatory realism," inspired by Wheeler's "participatory universe" (1994), was suggested by Fuchs (Fuchs 2017).

3 Observations

Two observations can be made in the light of Table 7.1.

Observation one. The extended belief that both types of interpretations "yield the same empirical consequences" and therefore "the choice between them cannot be made simply on purely physical grounds but it requires an act of metaphysical judgement" (Polkinghorne 2014) is arguably wrong. The two types are so different that it is very unlikely that distinguishing between them is forever out of reach of scientific method. In fact, by making some reasonable assumptions it can be shown (Cabello et al. 2016) that type-I interpretations must have more strange properties than those suggested by previous approaches (Bell 1964; Kochen and Specker 1967; Pussey, Barrett, and Rudolph 2012).

Observation two. The proposed principles for reconstructing quantum theory (Hardy 2001, 2011; Chiribella, D'Ariano, and Perinotti 2011; Dakić and Brukner 2011; Masanes and Müller 2011; Cabello 2013; Barnum, Müller, and Ududec 2014; Chiribella and Yuan 2014) are neutral with respect to interpretations. This may be a drawback. Although these approaches can lead to the mathematical structure of the theory, the need to remain neutral may be an obstacle for identifying *physical* principles. This becomes evident when the proposed principles are examined in the light of some specific interpretations, e.g., QBism (Fuchs 2010; Fuchs and Schack 2013; Fuchs, Mermin, and Schack 2014). Then, not all of them are equally important: some simply follow from the assumed interpretational framework; only a few of them give some physical insight. This suggests that there may be a bonus in non-neutral reconstructions of quantum theory. This will be developed elsewhere.

4 Note

This table (a map of madness) and these observations should be taken as motivations for further work. The fact that they may be controversial by themselves and the interest showed by many colleagues justify presenting them separately from any of these works.

Acknowledgments

This work was supported by the FQXi Large Grant project "The Nature of Information in Sequential Quantum Measurements" and project FIS2014-60843-P (MINECO, Spain) with FEDER funds. I thank D. Z. Albert, M. Araújo, L. E. Ballentine, H. R. Brown, Č. Brukner, J. Bub, G. Chiribella, D. Dieks, C. A. Fuchs, R. B. Griffiths, M. Kleinmann, J.-Å. Larsson, M. Leifer, O. Lombardi, N. D. Mermin, M. P. Müller, R. Schack, C. Timpson, L. Vaidman, D. Wallace, A. G. White, and A. Zeilinger for conversations and suggestions for improving the table.

References

Ballentine, L. E. (1970). "The Statistical Interpretation of Quantum Mechanics." *Reviews of Modern Physics*, **42**: 358–381.

Barnum, H., Müller, M. P., and Ududec, C. (2014). "Higher-Order Interference and Single-System Postulates Characterizing Quantum Theory." *New Journal of Physics*, **16**: 123029.

Belinfante, F. J. (1973). *A Survey of Hidden-Variables Theories*. New York: Pergamon Press.

Bell, J. S. (1964). "On the Einstein-Podolsky-Rosen Paradox." *Physics*, **1**: 195–200.

Bell, J. S. (1976). "The Theory of Local Beables." *Epistemological Letters*, **9**. Reprinted in (1985) *Dialectica*, **39**: 85–96.

Bohm, D. (1952). "A Suggested Interpretation of the Quantum Theory in Terms of Hidden Variables. I & II." *Physical Review*, **85**: 166–193.

Bohr, N. (1998). *The Philosophical Writings of Niels Bohr*. J. Faye and H. J. Folse (eds.). Woodbridge, CT: Ox Bow Press.

Brukner, Č. (2016). "On the Quantum Measurement Problem." Pp. 95–117 in R. Bertlmann and A. Zeilinger (eds.), *Quantum UnSpeakables II: Half a Century of Bell's Theorem*. Switzerland: Springer.

Bub, J. (1997). *Interpreting the Quantum World*. Cambridge: Cambridge University Press.

Cabello, A. (2013). "Simple Explanation of the Quantum Violation of a Fundamental Inequality." *Physical Review Letters*, **110**: 060402.

Cabello, A., Gu, M., Gühne, O., Larsson, J.-Å., and Wiesner, K. (2016). "Thermodynamical Cost of Some Interpretations of Quantum Theory." *Physical Review A*, **94**: 052127.

Chiribella, G., D'Ariano, G. M., and Perinotti, P. (2011). "Informational Derivation of Quantum Theory." *Physical Review A*, **84**: 012311.

Chiribella, G. and Yuan, X. (2014). "Measurement Sharpness Cuts Nonlocality and Contextuality in Every Physical Theory." arXiv:1404.3348.

Dakić, B. and Brukner, Č. (2011). "Quantum Theory and Beyond: Is Entanglement Special?" Pp. 365–392 in H. Halvorson (ed.), *Deep Beauty. Understanding the Quantum World through Mathematical Innovation*. New York: Cambridge University Press.

Dickson, W. M. (1998). *Quantum Chance and Nonlocality: Probability and Nonlocality in the Interpretation of Quantum Mechanics*. Cambridge: Cambridge University Press.

Einstein, A. (1936). "Physics and Reality." *Journal of the Franklin Institute*, **221**: 349–382.

Everett III, H. (1957). "'Relative State' Formulation of Quantum Mechanics." *Reviews of Modern Physics*, **29**: 454–462.

Faye, J. (2014). "Copenhagen Interpretation of Quantum Mechanics." In E. N. Zalta (ed.), *Stanford Encyclopedia of Philosophy* (http://plato.stanford.edu). Stanford, CA: Stanford University.

Fuchs, C. A. (2010). "QBism, the Perimeter of Quantum Bayesianism." arXiv:1003.5209.

Fuchs, C. A. (2017). "On Participatory Realism." Pp. 113–134 in I. T. Durham and D. Rickles (eds.), *Information and Interaction: Eddington, Wheeler, and the Limits of Knowledge*. Switzerland: Springer.

Fuchs, C. A., Mermin, N. D., and Schack, R. (2014). "An Introduction to QBism with an Application to the Locality of Quantum Mechanics." *American Journal of Physics*, **82**: 749–754.

Fuchs, C. A. and Peres, A. (2000). "Quantum Theory Needs no 'Interpretation.'" *Physics Today*, **53**: 70–71.

Fuchs, C. A. and Schack, R. (2013). "Quantum-Bayesian Coherence." *Reviews of Modern Physics*, **85**: 1693–1714.

Ghirardi, G. C. (2011). "Collapse Theories." In E. N. Zalta (ed.), *Stanford Encyclopedia of Philosophy* (http://plato.stanford.edu). Stanford, CA: Stanford University.

Ghirardi, G. C., Rimini, A., and Weber, T. (1986). "Unified Dynamics for Microscopic and Macroscopic Systems." *Physical Review D*, **34**: 470–491.

Gisin, N. (2012). "Non-realism: Deep Thought or a Soft Option?" *Foundations of Physics*, **42**: 80–85.

Goldstein, S. (2013). "Bohmian Mechanics." In E. N. Zalta (ed.), *Stanford Encyclopedia of Philosophy* (http://plato.stanford.edu). Stanford, CA: Stanford University.

Griffiths, R. B. (1984). "Consistent Histories and the Interpretation of Quantum Mechanics." *Journal of Statistical Physics*, **36**: 219–272.

Griffiths, R. B. (2014). "The Consistent Histories Approach to Quantum Mechanics." In E. N. Zalta (ed.), *Stanford Encyclopedia of Philosophy* (http://plato.stanford.edu). Stanford, CA: Stanford University.

Hardy, L. (2001). "Quantum Theory from Five Reasonable Axioms." quant-ph/0101012.

Hardy, L. (2011). "Reformulating and Reconstructing Quantum Theory." arXiv:1104.2066.

Harrigan, N. and Spekkens, R. W. (2010). "Einstein, Incompleteness, and the Epistemic View of Quantum States." *Foundations of Physics*, **40**: 125–157.

Hensen, B. et al. (2015). "Loophole-Free Bell Inequality Violation Using Electron Spins Separated by 1.3 Kilometres." *Nature*, **526**: 682–686.

Jammer, M. (1974). *The Philosophy of Quantum Mechanics: The Interpretations of Quantum Mechanics in Historical Perspective*. New York: Wiley.

Kochen, S. (1985). "A New Interpretation of Quantum Mechanics." Pp. 151–169 in P. J. Lahti and P. Mittelstaedt (eds.), *Symposium on the Foundations of Modern Physics: 50 Years of the Einstein-Podolsky-Rosen Experiment*. Singapore: World Scientific.

Kochen, S. and Specker, E. P. (1967). "The Problem of Hidden Variables in Quantum Mechanics." *Journal of Mathematical Mechanics*, **17**: 59–87.

Leifer, M. S. (2014). "Is the Wavefunction Real?" Unpublished talk at the *12th Biennial IQSA Meeting Quantum Structures* (Olomouc, Czech Republic).

Lombardi, O. and Dieks, D. (2012). "Modal Interpretations of Quantum Mechanics." In E. N. Zalta (ed.), *Stanford Encyclopedia of Philosophy* (http://plato.stanford.edu). Stanford, CA: Stanford University.

Masanes, L. and Müller, M. P. (2011). "A Derivation of Quantum Theory from Physical Requirements." *New Journal of Physics*, **13**: 063001.

Mermin, N. D. (2012). "Quantum Mechanics: Fixing the Shifty Split." *Physics Today*, **65**: 8–10.

Polkinghorne, J. (2014). "Physics and Theology." *Europhysics News*, **45**: 28–31.

Pusey, M. F., Barrett, J., and Rudolph, T. (2012). "On the Reality of the Quantum State." *Nature Physics*, **8**: 475–478.

Rovelli, C. (1996). "Relational Quantum Mechanics." *International Journal of Theoretical Physics*, **35**: 1637–1678.

Spekkens, R. W. (2007). "Evidence for the Epistemic View of Quantum States: A Toy Theory." *Physical Review A*, **75**: 032110.

Vaidman, L. (2015). "Many-Worlds Interpretation of Quantum Mechanics." In E. N. Zalta (ed.), *Stanford Encyclopedia of Philosophy* (http://plato.stanford.edu). Stanford, CA: Stanford University.

Van Fraassen, B. C. (1972). "A Formal Approach to the Philosophy of Science." Pp. 303–366 in R. Colodny (ed.), *Paradigms and Paradoxes: The Philosophical Challenge of the Quantum Domain*. Pittsburgh, PA: University of Pittsburgh Press.

Wheeler, J. A. (1983). "Law without Law." Pp. 182–213 in J. A. Wheeler and W. H. Zurek (eds.), *Quantum Theory and Measurement*. Princeton, NJ: Princeton University Press.

Wheeler, J. A. (1994). *At Home in the Universe*. Woodbury-New York: American Institute of Physics Press.

Zeilinger, A. (1999). "A Foundational Principle for Quantum Mechanics." *Foundations of Physics*, **29**: 631–643.

Zeilinger, A. (2005). "The Message of the Quantum." *Nature*, **438**: 743.

Part III

Probability, Correlations, and Information

8

On the Tension between Ontology and Epistemology in Quantum Probabilities

AMIT HAGAR

1 Introduction

Despite the so-called "quantum revolution," determinism is still the working hypothesis of modern physics. As far as our physical theories are concerned, the mathematical requirement for existence and uniqueness of solutions to the initial conditions, irrespective of the dynamical model that is used to describe the phenomena, is, and has been since the inception of mathematical physics, the first consistency test that such a model must pass. This is true even in what is commonly believed to be the exemplar of indeterminism, namely, quantum theory. In the nonrelativistic case, Schrödinger's equation maps every quantum state to one and only one quantum state, and this formal requirement holds regardless of the interpretational question of what this state actually means, a question that is more often than not delegated to the philosophers; in the relativistic case, the same formal requirement holds for Dirac's equation, and any divergences or singularities are further eliminated with the help of renormalization techniques. Quantum theories, relativistic or not, as far as their *dynamics* are concerned, are thus completely deterministic.

The origin of quantum *in*determinism, on the other hand, is still under dispute: most physicists would point at the measurement process as the seat thereof, yet, alas, there is still no consensus on how to formally incorporate the measurement process into quantum dynamics, and the specter of the disunity (von Neumann 1932) between "process 1" (the unitary, linear, and deterministic Schrödinger's equation) and "process 2" (the non-unitary and non-deterministic measurement process) still haunts us today.

In such a deterministic dynamical setting, a complete specification of the state of the system at one time – together with the dynamics – uniquely determines the state of that system at any other times. An observer, say a demon, who could attain a complete knowledge of the state of the system and its dynamics, would have no use for probabilities. As already noted by Laplace (1902), for such a demon, "nothing would be uncertain, and the future just like the past would be present before its eyes." Note, however, that these two ingredients, namely, complete knowledge of the initial conditions and deterministic dynamics, may be necessary but are not sufficient for complete predictability, for even Laplace's demon would need to consult an "oracle" if the question posed to him involved quantification over unbounded time (Pitowsky 1996).

In modern terms, complete knowledge means infinite precision, which is impossible in practice; unlike demons such as Laplace's, human beings are doomed to employ only finite precision in any physical measurement. A physicist who operates in this context, whose resolution power is less than optimal, is thus bound to make probabilistic statements, and, given the deterministic character of her dynamical models, such statements are commonly understood as subjective (Lewis 1986); the inability to predict an outcome exactly (with probability 1) is predicated on the notion of ignorance, or incomplete knowledge of the observer.

This subjective interpretation of probability is natural in the context of classical statistical mechanics (SM henceforth), where a physical state is represented as a point on phase space, and the dynamics is a trajectory in that space, or in the context of Bohmian mechanics, where the state and the history are still a point and a trajectory, respectively (but phase space is replaced with a configuration space and the dynamics is augmented with the quantum potential). Recently it has been suggested as a viable option also in the context of orthodox nonrelativistic quantum mechanics (QM henceforth), where the state is represented as a ray in the Hilbert space, and the dynamics is a unitary transformation, i.e., a rotation, in that space (e.g., Caves, Fuchs, and Schack 2002). All three cases share a strictly deterministic dynamics, but differ on the representation of the physical state and on the formal representation of probabilities: in classical SM or Bohmian mechanics, probabilities are subsets of phase space (or configuration space) obeying a Boolean structure (Goldstein, Dürr, and Zanghi 1992); in QM, they are angles between subspaces in the Hilbert space obeying a non-Boolean structure (Pitowsky 1989), whence non-locality, contextuality, and the violation of Bell's inequalities.

The subjective interpretation of probability in statistical physics appears to many inappropriate. The problem is not how lack of knowledge can bring about physical phenomena (Albert 2000: 64); it can't. Neither is it a problem about antirealism (Hagar 2003) or ontological vagueness (Bell 1990). Rather, the problem is that a subjective interpretation of probability in statistical physics, be it classical SM or QM, turns these theories into a type of statistical inference, and infects physics with a radical variant of dualism: while applied to physical systems, these theories become theories about epistemic judgments in the light of incomplete knowledge, and the probabilities therein do not represent or influence the physical situation, but only represent our state of mind, or degrees of belief (Frigg 2007; Uffink 2011).

That probability signifies subjective degree of belief and not an objective feature of the world was famously argued by de Finetti, who also insisted that "probability doesn't exist" (this striking motto is reported on the memorial tablet at his birthplace in Innsbruck, Austria). Here's a typical quote:

It does not help me at all to give the name probability to the limiting frequencies, or to any other objective entity, if the connection between these considerations and the subjective judgments which depend on them remain subjective. .... It is for these reasons that the theory of probability ought not be considered as auxiliary theory for the branches of science which have not yet discovered the deterministic mechanism that "must" exist; instead it ought to be regarded as constituting the logical premises of all reasoning by induction.

(de Finetti 1980: 110, 116)

There is a lot to admire in the attempt to impose rationality and structure on our beliefs – irrational and messy as they may seem – but such a view that confines the interpretation of probabilistic statements in statistical physics to the subjective realm, and regards probability theory as an extension of inductive logic, leads to some awkward consequences.

Think of a probabilistic statement such as "there is 50% chance of rain in the DC area tomorrow," based on a certain dynamical model of the weather. On the subjective account of probability, this statement can only mean that if one would have done a random sampling of the residents of the DC area, and asked them what is their belief about the weather tomorrow, around half of the subjects would answer they believed it would rain. So apart from telling us that we should expect a certain ratio of umbrellas at any given train in the DC subway tomorrow, we learn very little from this statement about the physical state of the weather. In other words, if probability and chance are *not* features of the physical world, then, contrary to what their practitioners and proprietors may think, the said dynamical models of the weather – indeed, statistical physics in general – do not describe this world, but rather our beliefs, or states of mind.

This tension between the deterministic character of the dynamics that underlies statistical physics and the subjective interpretation of probability as a measure of the observer's ignorance presents statistical physicists with the following dilemma, which for lack of a better name we shall here call the "Observer Dilemma":

OBSERVER DILEMMA: Either the observer is a part of the physical world, or she isn't. If she is, then her measure of ignorance may be part of the formalism which describes this world, but then must also signify a physical feature of this world; if she isn't, then her measure of ignorance need not represent any physical feature of the world, but also cannot be a part of a formalism which describes it.

The subjectivist, it seems, would like to have it both ways, and this explains the current schizophrenia in statistical physics regarding probability and the place of the observer in physics, manifest, e.g., in the famous confusion between Bohr and

Pauli on the notion of "detached observer" (Pauli 1994: 43–48; see Gieser 2005: 130–134): on one hand, we have a view of physics that includes a "participating observer," whose subjective perspective becomes an inevitable part of the objective physical description, and on the other hand, a view of subjective probability as signifying nothing but the degree of credence of a detached observer, which has no room in an objective physical description.

The purpose of this contribution is to offer an objective dynamical interpretation of probability that avoids the Observer Dilemma within a deterministic background. To do so we shall first look at a recent interpretation of QM, the Bayesian approach, that nicely exemplifies this dilemma. We shall then turn to a common alternative interpretation of probability in statistical physics, which claims to avoid the Observer Dilemma by detaching probability from the deterministic dynamics. Exposing this alternative's faults, we shall then offer a desiderata for any interpretation of probability that can escape the dilemma, and suggest a sketch of such an interpretation that goes back to Heisenberg's and Dirac's intuitions in 1926 on the meaning of quantum probabilities and the physical foundations of the uncertainty principle. The new interpretation, or so I shall argue, could serve as an objective alternative to the subjective approach to probability of the Bayesian approach.

2 What Is Wrong with Quantum Bayesianism?

Quantum Bayesianism is based on an epistemic attitude according to which the quantum state $|\psi\rangle$ (the amplitude mod square of which, $|\psi|^2$, is, after Max Born, interpreted as probability) does not represent a real physical state of a system, but instead supplies an observer with statistical information concerning all possible distributions of measurement results for all possible measurements. The probabilities computed by the standard Born rule are understood as probabilities of *finding* the system on measurement in some specific state, or, better yet, they represent betting rules for the observer on the possible results of future experiments. Applying von Neumann's projection postulate to the quantum state (or more generally applying Lüder's rule), under this account, is just an adjustment of subjective probabilities, conditionalizing on newly discovered results of measurement, i.e., it is merely a change in the observer's knowledge, or probability assignments. By contrast, the unitary and linear quantum mechanical dynamics describes the observer-independent and in this sense objective time-evolution of the quantum probabilities when no measurement takes place. Hence, in this approach measurements can be treated operationally as "black boxes" and require no further theoretical analysis.

Also here the tension between the deterministic (Schrödinger) dynamics and the probabilistic statements that the theory allows us to formulate is resolved by detaching the former from the latter. Also here there is a lot to admire in the attempt to endow our beliefs about the results of physical experiments with the kind of rational structure – which in the quantum case is also non-Boolean – that the Bayesian is so proud of. And also here the problem is that such an attempt turns out to be incomplete at best, or worse, inconsistent, since there is a certain type of experiments, possible *in principle*, for which this approach cannot give a unique prediction.

Think of the following thought experiment, which is a variant on Wigner's friend (1979). In this set-up, an observer A measures the z-spin of a spin-half particle P by means of a Stern-Gerlach apparatus (which, to keep things simple, we omit from our description). The quantum state of $P + A$ initially is

$$|\Psi_0\rangle = (\alpha|+_z\rangle + \beta|-_z\rangle)|\psi_0\rangle_A \tag{1}$$

where $|\alpha|^2 + |\beta|^2 = 1$ $(\alpha, \beta \neq 0)$, the kets $|\pm_z\rangle$ are the z-spin eigenstates, and $|\psi_0\rangle_A$ is the initial *ready* state of A. After the measurement, in a no-collapse theory, the quantum mechanical state of $P + A$ is the superposition:

$$|\Psi_1\rangle = \alpha|+_z\rangle|see\ up\rangle_A + \beta|-_z\rangle|see\ down\rangle_A \tag{2}$$

where $|see\ up\rangle_A$ and $|see\ down\rangle_A$ are, say, the brain states of A corresponding to her perceptions and memories of the two possible outcomes of the measurement. By contrast, in a collapse theory of the GRW kind, the state (2) is highly unstable (assuming that the chain of interactions leading to A's different memory states involves macroscopically distinguishable position states), so that by the time the measurement is complete this state collapses onto one of its components.

Consider now an observable $\hat{O}$ of the composite system $P + A$ of which the state (2) is an eigenstate with some definite eigenvalue, say $+1$. Observables like $\hat{O}$ are defined in the tensor product Hilbert space $\mathcal{H}_P \otimes \mathcal{H}_A$ unless super-selection rules are introduced. For our purposes, think of $\hat{O}$ as an observable that pertains to P's spin degree of freedom and the relevant degrees of freedom of A's sense organs, perceptions, memory, etc.

Suppose now that the composite system $P + A$ is completely isolated from the environment, and that a measurement of $\hat{O}$ is about to be carried out on $P + A$ immediately after the state (2) obtains. According to no-collapse QM, the measurement of $\hat{O}$, under these circumstances, is completely non-disturbing in the sense that after the measurement the state of $P + A$ remains precisely as in (2). One may think of $\hat{O}$ as an observable that is maximally sensitive to whether or not the interference terms between the different components of (2) exist. In other words, the measurement of $\hat{O}$ on $P + A$ if the state (2) is the true state of $P + A$ is a non-demolition measurement that, as it were, passively *verifies* whether or not $P + A$ is in fact in that state.

Note that $\hat{O}$ commutes neither with the z-spin nor with A's perceptions and memories of the outcomes of the z-spin measurement. This surely raises interesting questions about the status of the uncertainty relations in this set-up and about the reliability of A's memories of the outcome of her spin measurement in the event that A measures $\hat{O}$ just after her spin measurement. However, no matter what happens during the measurement of $\hat{O}$ (to A's memory of the outcome of her spin measurement, or to the z-spin values themselves), QM implies that the *correlations* between the z-spin of P and A's memories must remain exactly the same as they were before the $\hat{O}$-measurement. Moreover, in a no-collapse theory, the state of $P + A$ immediately after the $\hat{O}$-measurement will be, with complete certainty, just:

$$|\Psi_1\rangle = \alpha|+_z\rangle|see\ up\rangle_A|see\ \hat{O} = +1\rangle + \beta|-_z\rangle|see\ down\rangle_A|see\ \hat{O} = +1\rangle \tag{3}$$

where $|see\ \hat{O} = +1\rangle$ is the state corresponding to perceiving the result of the $\hat{O}$-measurement.

By contrast, in a collapse theory, the state of $P + A$ *immediately* after the $\hat{O}$-measurement will be given by *one* of the eigenstates of $\hat{O}$, where the probability that it will be state (2), and therefore that the outcome $\hat{O} = +1$ will be obtained, is only $|\alpha|^2$. Note that even if that outcome will obtain, the state (2) will extremely quickly collapse, again, onto one of the components of state (2) with probabilities that are given by $|\alpha|^2$ and $|\beta|^2$. So, in the GRW theory, the final value of the z-spin and the spin memory of A might be different before and after the $\hat{O}$-measurement.

Note further that we deliberately do not specify here who carries out the $\hat{O}$-measurement (i.e., in which degree of freedom the outcome $\hat{O} = +1$ is recorded). It may be carried out by A or by some other observer B external to A's laboratory. As can be seen from our notation in (3), we have implicitly assumed (for the sake of simplicity only) that the outcomes of A's spin measurement and of the $\hat{O}$-measurement are recorded in separate degrees of freedom. But in fact our argument here does not depend on this assumption. QM itself imposes no restrictions whatsoever on the way in which the outcomes of these measurements are recorded, except that they cannot be recorded *simultaneously* in the same degree of freedom (since $\sigma_z \otimes 1$ as well as A's memory observable are incompatible with $\hat{O}$). Moreover, QM (with or without collapse) imposes no further restrictions on the *identity* of the observers who may or may not carry out $\hat{O}$-type measurements (for an extended discussion of $\hat{O}$-type measurements and their implications, see Albert 1983; Aharonov and Albert 1981 use $\hat{O}$-type measurements in their discussion of the collapse of the quantum state in a relativistic setting).

To make things simple, let us suppose that the $\hat{O}$-measurement is to be carried out by the external observer B. But now we can ask A to give her *predictions* of the probabilities of the outcomes of the $\hat{O}$-measurement (clearly, in QM the quantum state assigned to a system is supposed to give the probabilities of the outcomes for *all* possible measurements). But here we encounter a problem in the Bayesian approach since it doesn't tell us on what quantum state A should base her predictions!

In order to calculate her expected probabilities, A may choose one of the following two options:

(a) Update her quantum state in accordance with the outcome of the spin measurement she actually observed, either $|see\ up\rangle_A$ or $|see\ down\rangle_A$. In this case, she would *collapse* the state (2) onto one of its components (i.e., spin+memory). Applying the Born rule to this state, she will predict that the result of the $\hat{O}$-measurement will be +1 with *probability* $|\alpha|^2$.
(b) Ignore the outcome of the spin measurement she actually observed, and conditionalize her probabilities on the *uncollapsed* state as in (2). In this case, since the state in (2) is an eigenstate of $\hat{O}$ with eigenvalue +1, she will predict that the result of the $\hat{O}$-measurement will be +1 with *certainty* (i.e., probability 1).

Note, however, that since the Bayesian does not know whether post-measurement states of the form (2) actually collapse, she does not know which of these two predictions is correct. But, surely, they cannot be *both* true, since they are inconsistent. And, even more embarrassing for the Bayesian, by carrying out a series of repeated $\hat{O}$-measurements on identically prepared systems, all in state (2), we can distinguish experimentally between these two predictions, (a) or (b), which means that quantum Bayesianism is an incomplete story that cannot account for experiments that quantum theory tells us are possible *in principle*.

A Bayesian would probably raise the objection that there is no harm in assigning different degrees of belief to the same situation when one has a different amount of knowledge to begin with, but the point to be made here is that, while there is plainly a fact of the matter about the correct result of the experiment, the Bayesian approach gives no plausible account of *which* option – (a) or (b) – is the correct one.

On the one hand, the full information about the lab available to A before the $\hat{O}$-measurement is given by the collapsed state, and this justifies option (a). On the other hand, if A knows QM under the Bayesian approach, then she believes that there is no real *collapse* of the state in measurement, and so her predictions ought to be guided in this case by the *uncollapsed* state as in (2). That is, on this view option (b) is justified. So, according to the Bayesian view, the two predictions seem to be on equal footing.

What our thought experiment shows is that once we treat measurements operationally, and probability as epistemic, or subjective, then the assignment of quantum states becomes (in some circumstances) ambiguous, and this ambiguity leads to incompatible probability assignments to measurement outcomes! And while there is nothing inconsistent in having different predictions based on different amounts of information, once we remind ourselves that these predictions actually refer to matters of fact in the world, we must, as it were, make up our mind which prediction is correct and which is not; they cannot be both true (for more details about this argument, its background, and the rebuttal of possible reactions thereto, see Hagar and Hemmo 2006).

It is important to note that this muddle is a direct result of the two ingredients that make up the Observer Dilemma: on the one hand, the Bayesian wants to keep the deterministic quantum dynamics intact; on the other hand, she insists on interpreting the probabilities that appear therein as purely subjective. That for a certain type of experiments the Bayesian cannot give a definite prediction only means that she is bound by our dilemma: by interpreting probability as purely subjective she cannot tie it back to the actual state of affairs in the world, a state of affairs presumably described by the quantum dynamics.

3 Typicality

One way to go about interpreting probability as an objective feature of the world, notwithstanding the deterministic character of the dynamics that presumably describe it, is, no surprise, to *detach* the probabilities from the dynamics. On this view probabilities arise from some initial distribution of events on the space of all possible events, or, in the dynamical context, from some distribution of initial conditions on phase space. Once endowed with this distribution at time t_0, the deterministic dynamics takes over, "carrying the probability" with the dynamical flow to any other time $t > t_0$. Probabilities thus enter once and only once in the beginning of the universe, and our current probabilistic statements reflect the initial distribution with which our universe came equipped. This view is known today as *typicality*:

The probabilities here are independent of the dynamical laws. If one is looking at a finite volume v of state space, there will almost always be multiple trajectories coming out of it and leading in different directions. Some will go up and some will go down, and to form expectations, one needs to know *how many* of a typical finite set of systems whose states begin in v go this way and *how many* go that. If one wants to know the probability that a system whose initial state falls in v is in v^* after an interval I, that is to say, she calculates the probability that a random pick from a typical ensemble of systems whose states begin in v, end in v^*. Dynamical laws that specify only what the physically *possible* trajectories will not yield these probabilities. And so where there is more than one possibility, dynamical laws will not specify how to divide opinion among them.

(Ismael 2009: 95)

Typicality as a view of objective deterministic chance has been defended both in the foundations of classical SM and in the context of Bohmian mechanics (e.g., Maudlin 2007). This seems natural, given the premise of the typicality approach, namely, the premise according to which the (deterministic) dynamics is *irrelevant* to the notion of probability; all that matters is the instantaneous representation of the state of the physical system, and, as a matter of fact, modulo the wave function, this representation of the physical state (as a point in phase space or as a point configuration space) is *identical* in classical SM and Bohmian mechanics.

Typicality claims tell us what *most* physical states are, or which dynamical evolutions (trajectories on phase space) are overwhelmingly more likely, by assigning measure 1 to the set of such states or the set of such dynamical evolutions. Examples are "most quantum states are mixed, i.e., entangled with their environment," or "most systems relax to thermodynamic equilibrium if left to themselves." Such claims make an analytical connection between a deterministic dynamics and a characterization of certain empirical distributions, hence can be interpreted as objective, having nothing to do with one's credence or state of knowledge. With this notion, or so the story goes, one can treat probabilistic statements in deterministic physical theories as arising from an objective state of affairs, and the theories that give rise to these statements as theories about the physical world, rather than theories about our state of mind.

How would the typicality approach to probability interpret statements such as "there is a 50% chance of rain in the DC area tomorrow"? On this approach, what this statement means is that, so far, in 50% of days macroscopically similar to today, it rained the next day in the DC area. Note that in order to render probability objective according to this view, an initial distribution is not sufficient; what is needed in addition is an agreed-upon separation between a micro-description and macro-description, and a specific choice of macro-states, usually on the basis of some agreed-upon notion of macroscopic similarity.

This objective view of probability, however, is still controversial, as it is saturated with many difficulties.

First, as its proponents admit (Goldstein 2012), the notion of typicality is too weak: a theorem saying that a condition is true of the vast majority of systems does not prove anything about a concrete given system. Next, the notion lacks logical closure: a pair of typical states is not necessarily a typical pair of states, which means that "being typical" is not an intrinsic property of an initial condition, not even for a single system, but depends on the relation between the state and other possible initial conditions (Pitowsky 2012). Possible ways around these difficulties have been suggested; Maudlin (2007: 287), for example, rejects the requirement to assign probabilities to single systems, and Pitowsky (2012) proposes to retain most of the advantages of typicality but to retreat to a full-fledged Lebesgue measure, with its combinatorial interpretation. But while these problems may be circumvented, there are three deeper lacunas underlying the notion of typicality that threatens the entire project.

The first point is that typicality claims depend on a specific choice of measure, usually the Lebesgue measure or any other measure absolutely continuous with it, and another choice of a uniform distribution relative to that measure. Yet classical phase space is an uncountable set, isomorphic to $\mathbb{R}^{6N}$ (where N is the number of degrees of freedom of the system at hand), and as such, there are infinitely many ways to endow it with a measure that can serve as a probability measure relative to which we can *count* points in that space. What criteria should one employ in order to choose such a measure and such a uniform distribution? And what justifies this choice of measure when an infinite number of possible measures are equally plausible?

A famous unsuccessful attempt to justify the choice of the Lebesgue measure and the uniform distribution relative to it in the foundations of SM is the ergodic approach (on this approach see, e.g., Sklar 1993 and reference therein). In this framework, one attempts to underwrite thermodynamics with mechanics by replacing the mechanical version of the law of approach to equilibrium with a probabilistic counterpart, derivable from the equations of motion. In special cases, namely when the dynamics is ergodic (i.e., when the only constant of motion is energy), equilibrium states are endowed with high probability, and so if a system started in a non-equilibrium state, it would, with certainty, approach equilibrium and remain there.

There are many objections to this approach (e.g., Bricmont 1996, Earman and Redei 1996), but the one most relevant to our story is that it is circular: the beautiful mathematical theorems, proven in this approach, apply only to a point-set of Lebesgue measure 1 on phase space, and yet the choice of the Lebesgue measure is exactly what these theorems are intended to justify in the first place.

Given this circularity, it seems best to adopt an empiricist stance (Hemmo and Shenker 2012), and follow a strictly empirical criterion: the choice of measure should have empirical significance, to the extent that it ought to be testable, as any other statement of physics. Once we narrow down the search for empirically adequate measures that yield probabilities that are close enough to the observed relative frequencies, we select a measure according to convenience.

This simple rule *may* yield the measure that is often used in the foundations of SM (namely, the Lebesgue measure relative to which every micro-state – a point in phase space – compatible with some macro-state – a delineated region of that space that represents some meaningful macroscopic physical magnitude – has equal weight), but the crucial point is that the arrow of justification for this choice starts in empirical adequacy: if we had evidence that the uniform distribution relative to that measure is not empirically adequate, we would have chosen a different measure and a different distribution! The moral, if there is one, is that questions about the choice of measure are empirical questions, not a priori ones (since choosing a uniform distribution often seems natural, the foregoing subtle point is quite general; so general that it reappears on many occasions *outside* SM. For a criticism of a similar attitude in the foundations of quantum gravity, see Hagar and Hemmo 2013).

The main difference between the typicality approach and the strictly empiricist one rests in the way both justify their choice in measure L. If one follows the latter, then one chooses measure L in accord with the observed relative frequencies of macro-states, that is, one makes an *additional* inductive leap, over and above the inductive leap that concerns the classical dynamical regularities. Since this choice is *based* on thermodynamic experience, it cannot *justify* our thermodynamic experience in a noncircular way (see the ergodic approach or the causal set approach). But if, on the other hand, one attempts to *explain* our experience using measure L (by saying, e.g., that it has some aesthetic dynamical virtues such as being conserved by the dynamics, or other a priori virtues), then one ends up in a non-defensible position that ultimately involves circular reasoning.

One might attempt to promote the initial distribution to the status of a new physical law (e.g., Loewer 2001), but such attempts are also doomed from the outset. A law of nature, however necessary and universal it may seem, presumably requires empirical evidence that can corroborate it, and thus can be overturned when contradicting evidence suddenly presents itself in domains yet to be tested. But according to the proponents of typicality, no amount of evidence can lead us to modify the initial distribution, and in particular no amount of *dynamical* evidence, i.e., competing dynamical laws, could do so, since this distribution is *independent* of any dynamical law, and is given to us only once.

The second problem is not unique to the typicality approach, but it is worth mentioning, as it emphasizes again the need for an empirical underpinning of the notion of probability. For even if we have established somehow that, relative to a preferred choice of measure, a certain set of states T_n is typical, i.e., its members are overwhelmingly more probable with respect to all possible states, what justifies the claim that we are likely to observe, or "pick up," members of that set T_n more often than members of the complement set T_{ab}? After all, the measure we have imposed on the space of all possible potential states $(T_n \cup T_{ab})$ need not dictate the measure we impose on the space of our actual observations; we could just as well toss a fair coin and endow observations of members of both T_n and T_{ab} with equal probability. This point is quite general and equally applies to QM: under the choice of the Haar measure (the analogue of the Lebesgue measure in SM), "most" states in the Hilbert space are mixed, hence entangled with their environment, but we can still realize (hence observe) pure states in the lab, at least to a certain extent. In what sense, then, are these states "rare" or abnormal?

The idea, known as *practical* (or *moral*) *certainty*, according to which events with small probability on the space of all *possible* events are correspondingly rare on the space of actual events hence are unlikely to occur, is prevalent in all accounts of probability, e.g., Bernoulli's, Laplace's, and Kolmogorov's, to name a few, and goes back to Leibniz (Hacking 1975: 146). But also here we have a problem in justifying the measure we impose on the space of actual events, which assigns high probability to those events with high measure on the space of possible events. As in the case of the choice of the measure discussed earlier, nothing but induction alone, i.e., empirical generalizations, allows us to assign low probability to the occurrence of "unlikely events" (unlikely in terms of their probability on phase space). For all we know, the next occurrence could be an occurrence of exactly such an "unlikely event," and we would have been wrong all along in our probability assignment. Some (e.g., Maudlin 2007: 277) call such a possibility "a cosmic run of bad luck," but no matter what one chooses to call this situation, the fact remains that nothing in the physical laws forbids it from holding in our world.

The twofold problem of justification of the measure is, on final account, another facet of the problem of induction (Pitowsky 1985: 234–238). On a strictly empiricist view, the effort to justify typicality claims with a priori considerations is just another (futile) attempt to give demonstrations to matters of fact, or to derive contingent conclusions from necessary truths. The point here is that there is no surrogate for experience in the empirical sciences, and that inductive reasoning is the best one can do in one's attempts to understand the world. A definition of objective probability should therefore be susceptible to such an empirical investigation and rely on as few a priori notions as possible.

The third and final problem stems from the fact that in the typicality approach, probabilities are postulated only once as a distribution on the initial conditions, and therefore the dynamics plays no role in their definition or calculation. While such an attitude might be defended on the basis of the common view about the incompatibility between determinism and objective probability, it also leads to some bizarre consequences. Think of a gas that is at t_0 compressed into a small volume of a cylinder with a valve. Once the valve is removed, the gas starts to expand until it fills up the cylinder. Mundane as this example is, the typicality approach gets it all wrong: according to this approach, in any given moment $t > t_0$ the probability that the gas fills the cylinder is practically 1, but of course, the process in which the gas fills the cylinder takes *time*; depending on this relaxation time, the prediction that one would find the gas in equilibrium in $t > t_0$ may be true or false. And yet according to the typicality approach, it is always true, no matter what t is.

4 Desiderata

Thus far we have seen that the common view, based as it is on the premise that if the dynamics is deterministic, probabilities must either be *subjective*, or, if construed objectively, they must be *detached* from the dynamics, is problematic and yields awkward situations: if one construes probability as an objective *static* feature of the world, one ends up in giving the wrong predictions in many cases of relaxation to thermal equilibrium; if one construes probability as a *subjective* feature, one ends up in giving ambiguous predictions in possible interference experiments even though there is a fact of the matter regarding the result thereof; and one is unable to account in any physical terms for a difference that one's own theory admits to exist between different physical states.

But why should one entertain this premise to begin with? There is no a priori reason to do so, and, as we have just seen, it leads to insurmountable problems. In the remainder of this chapter, I shall try to sketch what an alternative to this common view could look like.

The first conceptual shift required, notwithstanding determinism, is to construe the notion of physical probability as an objective *dynamical* feature of the world; objective physical probability should be seen as a transition probability from one state to another, and not merely as a distribution imposed on the initial conditions, "carried along" with the dynamics irrespective thereof.

Next, a lot of ink has been spilled on devising ingenious arguments, all aiming to demonstrate that probabilities, defined either as a measure of credulity or a measure of "size" on phase space, obey some a priori guidelines, be these rational, decision making, or aesthetic. But if physical probability is objective and dynamical, then any assumption about its origin must be *contingent*, derivable from some physical property that we can test, verify, and falsify. The sole criterion for such choice of measure is thus our experience, and the desire to reproduce the macroscopic, observable relative frequencies of everyday life.

Finally, the deterministic character of the dynamics, through the Observer Dilemma, constrains us to construe objective probability as a *physical* quantification of lack of knowledge, and not as a mere mental feature of the observer. One way to exorcise dualism and to go about fulfilling such a requirement is to regard "ignorance" or "lack of knowledge" as a measure of inaccuracy, an unavoidable consequence of limited resolution capability that can be quantified in precise physical and objective terms. This view of "lack of knowledge" as limited resolution allows us to interpret probabilities objectively as transition probabilities between any two measured physical states, while the dynamics between these states remain completely deterministic.

Probability, according to this new approach, should be seen as a physical, objective, and dynamical magnitude, which quantifies, in ways to be spelled out, the limited resolution power one has when one attempts to acquire information about the world via measurements. In what follows we shall first demonstrate the plausibility of this alternative in classical SM, and hint on how it can also be applied to nonrelativistic QM.

5 Probability as a Dynamical Concept

Classical statistical mechanics is done on phase space, a multidimensional space spanned by the positions and momenta of the particles that the physical system consists of. The *micro*-state of the physical system in any given moment is a point on that space. The (actual and possible) histories, or time-evolutions, of that system are trajectories on that space. At the very least, one principle constrains these histories, and it represents the idea of Laplacian determinism, namely, existence and uniqueness of solutions to the dynamical equations. On phase space it is manifest (i) in the requirement that any two trajectories or histories never meet, and (ii) in the requirement (known as Liouville's theorem) that throughout the history of the system, the region in phase space occupied by its possible micro-states, which are initially compatible with some (in general macroscopic) property, may change its shape, but not its volume.

Following a recent fresh outlook on the foundations of SM, let's call such a region, i.e., the set of micro-states that are initially compatible with some macroscopic property, "*a dynamical blob*" (see Hemmo and Shenker 2012). This shift in description from a point into a set of points, or a region, signifies the fact that unlike Maxwell's demon, we as human observers lack the resolution power to discriminate between possible micro-states, all of which are compatible with some macroscopic property. When these properties happen to designate meaningful physical magnitudes for creatures like us, given our correlations with the observed system, they are called '*macro-states*', and the remarkable *contingent* fact is that their behavior, governed by the underlying dynamics and by the way we correlate with physical systems, shows the kind of macroscopic regularity that thermodynamics describes.

The departure from the common view starts with the subtle distinction, made by Hemmo and Shenker (2012), between the dynamical blob and the partitioning of phase space into macro-states, and results in the *independence* of these two basic notions: a dynamical blob may start as a macro-state, but during its time-evolution it may change its shape, so that different points in it may end up in different macro-states. After all, the partitioning of the accessible region in phase space into macro-states depends only on the correlations between the observer and the system, and is stationary.

How can one define probability as an objective, dynamical feature in this context? If the dynamics is deterministic, probability can arise, or so the story goes, only from our ignorance of the exact state of the system. In our construction, what determines the transition probability of a physical system from one macro-state to another is the partial overlap between the dynamical blob and the macro-state:

$$P\left([M_1]_{t_2}/[M_0]_{t_1}\right) = \mu\left(B_{t_2} \cap [M_1]\right) \tag{4}$$

This means that the probability that a system that starts at a macro-state $[M_0]$ at time t_1 will end in a macro-state $[M_1]$ at time t_2 is given by the relative size μ of the dynamical blob B_{t_2} which overlaps with the macro-state $[M_1]$. Note that there is nothing subjective in this kind of transition probability. "Ignorance" here simply means lack of resolution power, which is expressed by the relation between dynamical blobs and macro-states, both of which are objective features of the physical world. Note also that in this construction, *entropy* and *probability* are two different concepts. The first, following Boltzmann, is a measure of the macro-state size; the second is the measure of the partial overlap of the dynamical blob and the macro-state. The first measure is chosen based on *empirical* considerations relevant to thermodynamics; the second is chosen on the *empirical* basis of observed relative frequencies. It *may* happen that once the first measure is chosen, it will turn out to be the same as the second measure, but there is no reason to require it. Nevertheless, *as a matter of fact*, empirical generalizations tell us that it is highly *probable* for entropy to obey the laws of thermodynamics, such as the second law or the approach to equilibrium.

In this context, the interpretation of probability as a measure of resolution power becomes evident: in thermodynamics entropy designates the degree to which energy is exploitable to produce work; the higher the entropy, the less the energy is exploitable. Exploitability means degree of control, which in mechanics translates into resolution power, or degree of precision; one has less control on the actual micro-state of one's system when that actual micro-state belongs to a macro-state of a larger size. Similarly, if probability is defined dynamically as a transition of the micro-state from one macro-state to another, a smaller probability means a less accurate resolution, in which μ, the overlap between the dynamical blob and the macro-state, is smaller.

The translation of this view of probability as a dynamical feature, a measure of inaccuracy of measurement in the transition from one state to another, to the quantum domain, may seem heretical. After all, there is a wide consensus that an unbridgeable metaphysical gap exists between the quantum and the classical; that an ignorance interpretation of probability in the quantum world has been debunked by the violations of Bell's inequalities, and that there is no point in asking what is

                                    Amit Hagar

the physical meaning of the quantum state, hence the shift to quantum information. But, as we shall now see, this heretical translation is exactly what is needed in order to escape the Observer Dilemma, and to endow quantum probabilities with objective, physical meaning.

As a matter of historical fact, a similar intuition was shared by Heisenberg and Dirac around 1926, during their attempt to supply an alternative to Schrödinger's wave mechanics, an attempt that led to the famous uncertainty relations.

Historically, the motivation for Heisenberg's paper on the uncertainty relations was to establish the consistency of his matrix mechanics with the experimental data, data that Schrödinger's wave mechanics represented by means of continuously evolving causal processes in space and time. Both Heisenberg and Dirac were thus concerned as early as the fall of 1926 with the question of whether the formalism of the newly constructed quantum theory allowed for the position of a particle and its velocity to be determinable in a given moment in time (Heisenberg to Pauli on November 15, 1926, reprinted in Pauli 1979: 354–355; Heisenberg to Peierls on June 30, 1930, reprinted in Pauli 1985; and Dirac 1927: 622–624). Their negative conclusion, enforced by intuitions they shared about spatial discreteness and non-commutative geometry, has led to the common view of the uncertainty principle as a restrictive empirical principle, call it (U_1), that states: *it is impossible to measure simultaneously position and momentum.*

Heisenberg's presentation of the uncertainty relation was purely qualitative, but a complete mathematical formulation thereof soon appeared (Kennard 1927), and was generalized by Robertson (1929):

$$\Delta_\psi A \; \Delta_\psi B \geq \frac{1}{2} |\langle [A, B] \rangle_\psi| \qquad (5)$$

Here A and B are any self adjoint operators, $\langle \cdot \rangle_\psi = \langle \psi| \cdot |\psi\rangle$ denotes the expectation value in state $|\psi\rangle$, Δ_ψ is the standard deviation of the observable in the state vector $|\psi\rangle$, and $[A, B] := AB - BA$ denotes the commutator. This generalization – together with the one given by Schrödinger (1930) – established the uncertainty relations as theorems of QM (for all normalized state vectors $|\psi\rangle$ on the Hilbert space), but it differed in its status and its intended role from Heisenberg's original presentation. Rather than statements about empirical facts concerning measurements, the uncertainty relations were now considered as statements about the spreads of the probability distributions of the several physical quantities arising from the same state. For example, the uncertainty relation between the position and momentum of a system may be understood as the statement, call it (U_2): *in any quantum state the position and momentum distributions spread cannot both be arbitrarily narrow.* Inequality (5) is an example of such a relation in which the standard deviation is employed as a measure of spread. A recent analysis has exposed the shortcomings of this representation, offering more suitable measures of spread (Hilgevoord and Uffink 1983, 1989; Uffink and Hilgevoord 1985).

Part of Heisenberg's reasoning in defending (U_1) was the attempt to show, by way of his famous Gamma-ray microscope thought experiment, that the inability of QM to precisely discern both momentum and position of a particle is consistent with the optimum accuracy obtainable in experimental measurements. On this view, the reason for the uncertainty relations was to be found in the semi-classical analysis of the measurement interaction. This interaction, when combined with Einstein-de Broglie relations (which relate wavelength to momentum via Planck's constant), involves a scattering process between a photon and a material body, say, an electron, and results in the latter's changing its momentum discontinuously, in a manner inversely proportional to the wavelength of the photon. This reasoning has led to yet another formulation of the uncertainty principle, call it (U_3): *it is impossible to measure position of a particle without subsequently disturbing its momentum, and vice versa.*

The history of the reasoning behind (U_3) is well known: Bohr, for one, saw it as yet another indication that the uncertainty relations in particular, and QM in general, are ultimately based on wave-particle duality, or complementarity (Jammer 1974), and a debate between him and Heisenberg ensued on the precedence of these principles as foundational principles of QM over the bare mathematical formalism. Setting aside this debate, which belongs to the folklore of the Copenhagen interpretation (e.g., Beller 1999: chapter 11), two issues here remain central to this chapter. First, it is noteworthy that while Heisenberg was concerned with measurement "error" or "disturbance," the connections of these with the notion of standard deviation is not straightforward (Hilgevoord and Uffink 1988b), and so, without further qualifications, (U_3) does not follow from, or entails, (U_2). Second, and even more important, Heisenberg's microscope thought experiment, and the "disturbance" view associated with it, consist of an alternative to Born's statistical interpretation of the state vector (Born 1926), or to Jordan's version of quantum theory (Duncan and Janssen 2013), as yet another context where probabilities enter into the formalism of QM. At that time, recall, the founding fathers of QM were indecisive with respect to the statistical character of the theory. Jordan and Born, for example, kept open the possibility that nature is intrinsically indeterministic. In contrast, both Heisenberg and Dirac held to the belief that nature's dynamics is deterministic. But while Born's statistical interpretation designated the square of the amplitude of a *stationary* state vector as the probability that the system is in that state, Dirac (1927: 641) and Heisenberg (1983) identified the statistical element of QM only in the context of experiments, or observations, where transitions from one state to another occur:

One can, like Jordan, say that the laws of nature are statistical. But one can also, and that to me seems considerably more profound, say with Dirac that all statistics are brought in only through our experiments. That we do not know at which position the electron will be the moment of our experiment, is, in a manner of speaking, only because we do not know the phases, if we do know the energy ... and in *this* respect the classical theory would be no different. That we *cannot* come to know the phases without ... destroying the atom is characteristic of QM.

(Heisenberg to Pauli, February 23, 1927, in Pauli 1979: 377; emphasis in the original)

This idea is also manifest in Heisenberg's remark that "the uncertainty relation doesn't refer to the past" (1930: 20), that is, the uncertainty relation should be interpreted as the inability to measure both the *current* position and the *future* momentum of a particle. The problem with this view was that, similarly to (U$_3$), without an objective alternative thereto, the transition probabilities it involves were commonly interpreted as purely subjective and epistemic, representing as they were the lack of knowledge of the observer, whose measurement "disturbs" a preexisting value (von Liechtenstern 1955). Yet the existence of such definite, classical values, which remain unknown due to dynamical limits of experiments (or due to the quantum nature of the measurement process), is usually (e.g., Brown and Redhead 1981) taken to be in conflict with the confirmed violations of Bell's inequalities (Aspect et al. 1982) and with the famous "no hidden variables" proofs (Kochen and Specker 1967). In other words, the idea that the measurement "disturbs" the quantum state is rejected by the common belief that a deep metaphysical difference exists between quantum and classical mechanics.

And yet, since the meaning of objective physical probabilities, or so I shall argue, in both SM and QM is the same, namely, it quantifies the lack of resolution power in a physical measurement, then there is no point in looking for such a metaphysical difference between the two theories. But that no qualitative *metaphysical* difference exists doesn't mean no difference exists at all. And so in order to defend an objective, dynamical transition probability in nonrelativistic QM within a deterministic framework (i.e., *without* collapse), we need to establish some non-metaphysical difference between the two theories, at least with respect to the probabilities they employ. In particular, we need to establish a quantitative difference, not a qualitative one.

And as it turns out, Heisenberg and Dirac have already touched upon such a difference back in 1926.

6 Discreteness

The speculation that QM in general, and the uncertainty principle in particular, militates against the spatial continuum was voiced early on by Einstein (Stachel

1993; Hagar 2008), and was toyed with – to varying degrees of seriousness – by at least two other major players in the history of quantum theory, namely, Dirac and Heisenberg. Among the two, Heisenberg was the most vocal, and is also well known for his lifelong search for minimal length (Carazza and Kragh 1995; Kragh 1995). In a letter to Pauli from November 15, 1926, he writes:

[I]f it ends up that space-time is somehow discontinuous, then it will really be very satisfying that it wouldn't make sense to talk about, e.g., velocity $\dot{x}$ [the first derivative of position–AH] at a certain point x. Because in order to define velocity, one needs at least 2 points, which, in a discontinuum-world, just cannot lie infinitely close [*benachbart*].

(Heisenberg to Pauli, November 15, 1926; in Pauli 1979: 354–355)

Dirac was less explicit, but it is by now well documented that his work was influenced by ideas consistent with the hypothesis of finite spatial resolution, e.g., finitist and operational ideas such as Whitehead's method of extensive abstraction, or the geometrical interpretation of non-commutativity (Darrigol 1992). The latter was for Dirac the most distinctive feature of the new quantum theory that separated it from the classical one (e.g., Dirac 1929). Dirac was also well aware of the methodological problems that accompanied the notion of a finite spatial resolution (e.g., the loss of relativistic causality at short distances), but wasn't as worried about these problems as his contemporaries were. This attitude is evident in his solution to the problem of the self-energy of the classical electron, which introduces a finite electron radius (1938), in his interpretation of this cutoff as an inherent feature of space-time (and not just as a limit on the theory's applicability), and in his famous and persistent criticism of the renormalization method (1963).

In the late 1920s, only a few physicists, motivated by finitist, operationalist, and relationalist views, were concerned with the limitations on spatial resolution in a single experiment, and relied on the Compton effect to exemplify their point, by drawing on Heisenberg's Gamma-ray microscope thought experiment (e.g., Ruark 1928; Wataghin 1930; for a comprehensive account, see Hagar 2014: chapter 4, and references therein). As such, these ideas succumb to the criticism of the epistemic interpretation of the uncertainty relations, but they also contain the basic element for its alternative, namely, the assumption that in any physical interaction, the momentum transfer between any two physical entities is finite and bounded from above. This cutoff on momentum that limits the spatial resolution of any position measurement was central to the development of the notion of fundamental length in field theories, and is the key to understanding of the uncertainty relations in a non-epistemic, physical, and objective sense.

The idea to introduce a cutoff to the momentum transfer in particle collision was first suggested by Wataghin (1934a, 1934b), and later elaborated by March (1951).

Both physicists were trying to eliminate the divergences that plagued the newly born quantum electrodynamics (QED), and in particular the self-energy of the electron, which originated from the notion of a point-coupling, due to which the theory was forced to admit waves of arbitrary small wavelengths (von Neumann 1932). One way to limit the number of waves was to remove the effectiveness of those waves with a frequency exceeding a certain limit. To this end, what served as an appropriate principle was the introduction of a new universal constant l_0, that, analogously to $\hbar$ and c, limits *in principle* the ability of observation, to the extent that one could not ascertain the position of a particle at rest with a greater accuracy than with a possible error l_0.

The fundamental length l_0 was introduced relativistically into the interaction Hamiltonian between the field and the particle as the minimal radius of a sphere, below which no interaction takes place. The alteration of the interaction Hamiltonian was done in such a way that momentum was always conserved. Consequently, the corresponding matrix element for, e.g., an absorption of a photon by an electron that is initially at rest for wavelength $\lambda \gg l_0$ remained unchanged, but for $\lambda \ll l_0$ it was reduced to one-half the classical value. For emission, the result was that an electron at rest was incapable of transmitting photons with wavelength smaller than l_0.

From these corrections to the interaction Hamiltonian, March (1951: 277) concluded that it was impossible to distinguish the positions of two particles by means of a diffraction experiment performed with light rays, if the distance between them was less than l_0. In such an experiment, a radiation would be required with a wavelength the order of magnitude of which would correspond to the distance to be measured. The zero-point energy divergence was now eliminated because waves with $\lambda \ll l_0$ could not be reflected from the walls of the cavity in which the radiation is enclosed, as neither free nor fixed electrons were able to deflect a photon by a finite angle.

It is noteworthy that, at least during the 1930s, physicists who introduced minimal length into field theories with the method of a high momentum cutoff remained agnostic about the structure of space-time, and interpreted the inability to resolve distances smaller than the minimal length as the limit of applicability of the theory. Such agnosticism was natural given the problems one encountered when trying to translate the (upper) bound on momentum space into a (lower) bound on position space, problems that include, among other things, breakdown of relativistic causality and tension with Lorentz's invariance.

Apart from the reliance on set-ups such as Heisenberg's Gamma-ray thought experiment, the connection between the limit on spatial resolution and the "disturbance" view of the uncertainty principle is also apparent from the famous debate on the epistemic coherence of QED. In 1931, Landau and Peierls argued (1983) that the newly born QED was epistemically incoherent as it precluded its own verification by allowing singularities (situations in which the theory gave no predictions) even in single measurements of field quantities. As Bohr and Rosenfeld (1957) later showed, one could restore epistemic coherence to the theory and eliminate the singularities, if one accepted that physical magnitudes should be defined not at a dimensionless spatial point, but rather in a finite extended spatial volume. The same reasoning was generalized in the late 1930s (Bronstein 1936), and later during the 1950s (Mead 1964), to include measurements of the gravitational field, and is part of the prehistory of quantum gravity (see Hagar 2014: chapter 5).

This role of the limit on spatial resolution as eliminating singularities and restoring epistemic coherence to a given theory is quite general, but still goes widely unappreciated. Take, for instance, Bohmian mechanics, which reproduces Born's rule (hence the statistical predictions of nonrelativistic QM) under the assumption of uniform probability distribution relative to the Lebesgue measure imposed on the initial conditions in 3-space (Goldstein et al. 1992). A question arises – but is rarely asked – regarding the subset of these initial conditions situated on the surface whose width is of measure zero and that separates 3-space into two equal halves (I thank M. Hemmo and O. Shenker for this example). Bohmian mechanics seems to give no predictions about the results of any experiment done on a physical system whose initial state belongs to the aforementioned subset (the same situation can also arise in collapse interpretations of nonrelativistic QM). Clearly, if position cannot be resolved with actual infinite precision *in principle* in any measurement, then singularities are eliminated, and epistemic coherence is restored. (It is also noteworthy that if one could achieve such an infinite precision, one could also detect the preferred rest frame that underlies Bohmian mechanics, or the presumably instantaneous collapse process.) We shall return to this point in Section 9, as this generality – i.e., the fact that the epistemic coherence of any theory coincides with the limit on spatial resolution, regardless of the metaphysical view involved – is key to the alternative to the "disturbance" view of the uncertainty principle we shall now turn to.

7 Uncertainty

It is quite remarkable that the empirical status of the uncertainty relations is far from being experimentally confirmed even today (Busch, Heinoen, and Lahti 2007). Lacking an empirical underpinning, the standard view of the uncertainty relations sees them either as a mathematical theorem of the formalism of QM (Robertson 1929), or as a manifestation of Bohr's ideas on wave-particle duality and complementarity (Jammer 1974). Heisenberg's attempt to physically motivate the uncertainty relations with a thought experiment that depicted them as arising from a "disturbance" inherent to the act of measurement has been widely criticized as untenable (Brown and Redhead 1981), the common view being that this line of reasoning lends support to an epistemic, hidden-variables interpretation of QM (von Liechtenstern 1955), which stands in conflict with the confirmed violations of Bell's inequalities (Aspect et al. 1982 and with famous "no hidden variables" proofs).

Here we shall suggest a different physical underpinning of these relations that is immune to this criticism, and that instead relies on two simple assumptions:

(a) It is *in principle* impossible to perform a position measurement with infinite precision.
(b) Every measurement can be reduced to (or construed as) position measurement.

Assumption (a) follows from the hypothesis of finite nature (see, e.g., Feynman 1965: 57 or Fredkin 1990: 255). Note that, as a matter of fact, assumption (a) is consistent with any position measurement we have ever made. That assumption (b) is plausible follows from the fact that a well-known theory, namely Bohmian mechanics, is empirically indistinguishable from nonrelativistic QM. Relying on a stronger version of assumption (b) as it does, namely, that every physical property ontologically supervenes on position (here I adopt only the weaker, operationalist version of this assumption, while remaining agnostic about the ontology of properties other than position), Bohmian mechanics nevertheless reproduces the Born rule under an initial equilibrium assumption (i.e., a uniform probability distribution in position space). Consequently, any counterexample to assumption (b) would automatically be a counterexample to the applicability of nonrelativistic QM. Indeed, in his path integral formulation of QM, Feynman famously states that "a theory formulated in terms of position measurements is complete enough in principle to describe all phenomena" (Feynman and Hibbs 1965: 96). Bell is yet another physicist who promoted assumption (b) on many occasions (e.g., Bell 1987: 52–62).

To see how these two assumptions physically underpin the uncertainty principle, let's take the mathematical formulation thereof that dispenses with standard deviation as an inadequate measure of spread of the probability distribution, and replaces it with more appropriate measures of spread (see Hilgevoord and Uffink 1983, 1988b; Uffink and Hilgevoord 1985). Consistent with Heisenberg's intuition, in this formulation, the uncertainty relation represents the inability to measure both the *current* position and the *future* momentum of a particle. In the context of Heisenberg's microscope thought experiment it can be written as (Hilgevoord and Uffink 1990: 133):

$$w_x\, W_p \geq C \tag{6}$$

where C is a constant of order unity. Relation (6) connects the resolving power of the measuring apparatus (w_x) with the predictability of the momentum of the measured object after the measurement (W_p). Mathematically, the former magnitude is the width of the wave function of the state of the object in position space, which represents how far is that state from a spatially translated copy of itself (Hilgevoord and Uffink 1989). The latter magnitude is the overall width of the wave function of the state in momentum space; it represents how narrow is the probability distribution of momentum, i.e., how small is the inaccuracy in its prediction.

By assumption (a), the first magnitude (w_x) is finite; by assumptions (b) and (a) – since we actually measure position when we measure momentum – the second magnitude (W_p) can be construed as representing spatial resolution hence is also finite. We can physically justify relation (6) once we realize that the momentum measurement depends in its accuracy on the (finite) accuracy of the earlier position measurement, and vice versa. Note that, in accordance with our goal to propose an objective notion of uncertainty and probability, this underpinning also replaces Heisenberg's intuition about the uncertainty arising from "disturbance" – in this case, the disturbance of the momentum of the particle by the position measurement – with an objective, "hidden variable free" notion of uncertainty that arises from an inherent limitation on spatial resolution: the point is that nowhere does this view involve the actual (but presumably unknown) *state* of the particle; what is being "disturbed," or modified, is the consecutive *measurement*, whose accuracy is inversely proportional to the accuracy of its alternate measurement.

So much for non-commutativity. What about interference? For this consider the proverbial double-slit experiment, where interference is explicitly addressed. Bohr famously argued (1983: 25) that the attempt to discern which slit the particle went through "destroys" the interference pattern on the photographic plate. If one considers Bohr's set-up (Hilgevoord and Uffink 1990), one notes that the detection of the slit the particle went through relies on the recoil of the screen due to the passage of the particle in that slit. This recoil sets an upper bound on the initial knowledge of the momentum of the movable screen: for this recoil to be distinguishable, the initial momentum of the screen along the y-direction (the direction perpendicular to the flight of the particle) must be known with an uncertainty no bigger than the same momentum recoil.

Next, Bohr reasoned that this upper bound in momentum imposes, via the uncertainty relation, a lower bound on the overall width of the wave function in position space, which is greater than the width of the interference band, and that if one attempts to detect the path of the particle, the interference vanishes. While Bohr's reasoning was incorrect (the reasoning involved an inappropriate measure of probability spread, namely, the translation width of the screen in position space, which turns out to be insensitive to the width of the interference pattern; see Hilgevoord and Uffink 1990: 131), his conclusion was right! This conclusion, namely, that the conditions under which a measurement that allows the distinction of the two paths the particle could have gone through *exclude* the conditions under which interference can be observed, is demonstrated succinctly in Hilgevoord and Uffink (1990: 135–136, and 1988a), where the appropriate width (or measurement resolution) that relates the interference pattern to the distinguishability of the two momentum states of a movable screen (associated with the passage of the particle in one slit or the other) is the width of the *visibility* of the interference pattern (rather than the width of the interference bands that Bohr used in his reasoning). Thus, whenever these states can be completely distinguished, that is, if the slit through which the particles pass can be determined with certainty, the visibility of the interference pattern also vanishes (for an actual, and not only *Gedanken*, experiment of precisely this sort, see Buks et al. 1998).

Given assumption (b), both widths, the distinguishability of the momenta and the visibility of the interference pattern, physically represent spatial resolutions of position measurements. That a momentum measurement can be construed as a position measurement was argued previously; that a measurement of the visibility of the interference pattern can be so construed follows from a simple analysis of interferometry: there, *visibility* is defined as a function of the *intensity* of the beams, which, in turn, is a function of the *spatial displacement* of the mirrors that constitute the interferometer (Uffink 1985). As shown in Hilgevoord and Uffink (1990), the two resolutions are dependent since the visibility is proportional to the matrix element, which is a direct measure of the distinguishability of the momentum states of the screen.

One issue still requires careful attention, namely, how can we explain that, if for any pair of consecutive measurements the commutator fundamentally *never* vanishes, there are nevertheless many such pairs for which – at least *for all practical purposes* – the commutator *appears* to vanish?

8 Over-description

Assumption (a) may look at first far too radical, as its combination with (b) seems to make any pair of consecutive measurements strictly non-commuting. The key, however, to the reconciliation of the above physical underpinning of relation (6) with classical mechanics (where measurements always commute), or with the case of commuting operators in QM, i.e., where there are no interference terms, lies in the following operational interpretation of commutativity as a manifestation of the over-description of the discrete by the continuous. This over-description is germane to all our theories, QM included, that employ a mathematical machinery that has much more structure than is required to model physics. Consequently, a lot of effort is spent in disabling or reinterpreting these redundancies, so that the modeling can be done in spite of them.

A familiar example are Maxwell's equations: the state space for these equations is of cardinality $(\mathbb{R}^6)^{\mathbb{R}^3}$. To deal with this uncountable set of states, we concentrate on systems having very special properties – e.g., continuity, uniformity, locality, linearity, or reversibility – Maxwell's equations happen to have all of these properties at once (Toffoli 1984). These features allow us to disregard most of the infinities: continuity and linearity, for example, mean that a small perturbation in the system's initial state leads to a correspondingly small perturbation in its final state, so that we don't have to worry about capturing its state with infinite precision, albeit allowed by the mathematical structure of the theory, as any error in this description would be bounded.

I suggest to view commutativity as yet another example of such an over-description. In classical mechanics, it represents the coarse-grained character of classical spatial resolution, which is presumably infinite. Spatial measurements can be regarded, from a coarse-grained perspective, *as if* they are done with infinite resolution, and consequently one can make the inaccuracies thereof arbitrarily small. Mathematically this means that classical measurement errors, while possible, are just a practical nuisance that can be compensated for and eliminated by arbitrarily enlarging the scope of the physical description without cost. Operationally, commutativity ensues from the fact that infinite precision is allowed, and so fundamentally there is no bound on physical resources involved in the measurement process. Consecutive measurements in classical mechanics thus appear to be (and so can be described as) independent of each other's inaccuracies *for all practical purposes*.

In QM the situation is no different. Here commutativity represents the operational independence of any two measurements, or their compatibility. Contrary to the uncertainty relations, or, say, to the case of any two consecutive orthogonal spin measurements, in those cases where measurements commute, the accuracy (or lack thereof) in one does not depend on the accuracy (or lack thereof) in the other. From the operational perspective developed here, this independence can arise in two types of scenarios. First, the physical process of construing, or performing at least one of the measurements as a position measurement involves, as a matter of fact, correlations with more degrees of freedom, and leads, as in the classical case, to a similar spatial coarse-graining (think, e.g., of temperature measurement with an analog thermometer whose scale is too coarse to represent minute errors in position). Such coarse-graining – commonly referred to as "decoherence" – makes the measurements *practically* insensitive to the respective inaccuracies in the spatial resolution. Second, a temporal coarse-graining is also possible, where commutativity is achieved in the so-called "thermodynamic limit," i.e., in an extremely (ideally infinite) long measurement process, as in the quantum adiabatic theorem (Messiah 1961: 739–746), or in the so-called "protective measurement" scenarios (Aharonov, Anandan, and Vaidman 1993).

The crucial point is that in both types of coarse-graining, one can treat the inaccuracies in spatial resolution as if they are arbitrarily small, hence – *for all practical purposes* – one can describe the pair of consecutive measurements as commuting.

This reasoning suggests that while the previous physical underpinning of the uncertainty relations is indeed radical, there is a way to make it consistent with the classical or the quantum formalism, and with the predictions of both theories, at least in the nonrelativistic domain (the generalization to the relativistic domain may be more involved, but certainly not a priori impossible). The price, as hinted previously, is that nonrelativistic QM is now seen as a phenomenological, "effective" theory, whose mathematical structure (the Hilbert space), rather than a fundamental structure that requires "an interpretation," is actually a mere set of ideal analytical tools for computing the probabilities of future states of an underlying deterministic and discrete process, from the (inherently) limited information we can have about that process. We shall say more on the consequences of this view in the final section.

9 Disturbance

Heisenberg, we recall, attributed the non-commutativity in the uncertainty relations to the measurement act of the position of the particle and the "disturbance" incurred thereby to its subsequent momentum. This attribution was later criticized for resurrecting the specter of hidden variables. Lacking an alternative non-epistemic justification, the uncertainty relations remained a mathematical theorem, a by-product of the structure of the Hilbert space.

The finitist underpinning of the uncertainty relations (Section 7), together with the possibility proofs that reproduce the empirical content of nonrelativistic QM from an underlying discrete structure based on finite measurement resolution (Hagar and Sergioli 2014), can be considered precisely such a viable alternative, and can render quantum probabilities objective and non-epistemic.

This alternative *starts* from measurement outcomes and their finite resolution, and seeks to reconstruct non-commutativity (or the non-Boolean structure of quantum probability) from these outcomes. As such, it practically *moots* the question of hidden variables, and renders any other question regarding the "meaning of the wave function" a red herring, exchanging as it does the "disturbance" of the particle's state with a "disturbance" of the spatial resolution of the consecutive measurement, irrespective of the particle's state.

On the finitist view developed here, what matters is the state space of measurement outcomes, now taken as primitive, and not the metaphysics of the quantum state. As long as the non-commutativity between consecutive spatial measurements is preserved, the structure of this state space of measurement outcomes allows for non-Boolean probabilities (Pitowsky 1994). That such an operationalist view (which elevates finite-resolution measurement outcomes into the status of the basic building blocks of the theory) is agnostic of metaphysics is by no means an argument against it, as long as it succeeds in reproducing from these building blocks, under certain conditions, the probabilistic predictions of QM and the structure of the Hilbert space that encodes them (again, for a more detailed presentation of the conditions under which such a reproduction is possible, see Hagar and Sergioli 2014).

For this reason, the finitist is immune to the standard critique of the semi-classical analysis of the entire class of thought experiments that followed Bohr and Rosenfeld, which relied on Heisenberg's "disturbance" view. This criticism, namely, that the "disturbance" view assumes the very classical ontology it purports to deny, can be blocked by noting that the said "disturbance" is not a disturbance of the existing-yet-unknown state, but a modification of the spatial resolution of the consecutive (position) measurement. This shift in reference is what allows the finitist to replace the epistemic notion of probability with an objective alternative, in which probabilities are interpreted not as degrees of belief resulting from subjective lack of knowledge of an existing reality, but as genuine physical transition probabilities between any two measurement outcomes.

To repeat, this objective view of probability is completely agnostic with respect to metaphysical questions such as "does a physical system possesses a definite state when not measured?" The only thing that matters is that the probabilities in the model match the empirical frequencies we observe in the classical as well as in the quantum domains (depending as they do on the finite or infinite spatial resolutions that underlie the respective probability structures), and the difference between the two domains is seen as a difference in probability measure (Boolean vs. non-Boolean), rather than a difference in ontology (hidden variables vs. no hidden variables).

This reasoning also changes the way we think about quantum probabilities: on the view proposed here, and following Heisenberg's and Dirac's intuition from 1926, these now result from an inherent (as opposed to merely practical) lack of precision in measurement, and are dynamical transition probabilities between any two measurement outcomes, and as such qualify as an objective alternative to subjective view of nonrelativistic QM.

Concluding with a historical perspective, it is noteworthy that, at least until he was swayed by Bohr, Dirac presumably entertained a similar agnostic intuition (Darrigol 1992: 347), as he displayed no worries about the system's possibly having (at any given time) definite coordinates x and p, as long as these were represented as matrices and not as numbers. On this view, his theory of transformations just implied that, *given* non-commutativity, it was fundamentally impossible to predict unambiguously the state of the system at a subsequent time. Under the pressure of the Copenhagen school, Dirac was later persuaded to give up the so called "fiction" of a definite x and p, and in his later presentations of QM adopted the formal notion of "state vector" proposed by Weyl and von Neumann, the metaphysical interpretations of which have generated no end of trouble. If this chapter has made you just a little bit more of a Copenhagen disbeliever, dear reader, then I have earned my day's work.

Acknowledgments

I am grateful to Olival Freire Jr. from the Program of History of Physics at the Institute of Physics, UFBA, where this project was conceived, under the generous support of the Fulbright Mobility Award to Brazil. Thanks also to G. Ortiz (IU Physics) for discussion, and to Olimpia Lombardi for tolerating my heretical views.

References

Aharonov, Y. and Albert, D. (1981). "Can We Make Sense of the Measurement Process in Relativistic Quantum Mechanics?" *Physical Review D*, **24**: 359–370.

Aharonov, Y., Anandan, J., and Vaidman, L. (1993). "Meaning of the Wave Function." *Physical Review A*, **47**: 4616–4626.

Albert, D. (1983). "On Quantum Mechanical Automata." *Physics Letters A*, **98**: 249–252.

Albert, D. (2000). *Time and Chance*. Cambridge, MA: Harvard University Press.

Aspect, A., Grangier, P., and Roger G. (1982). "Experimental Realization of Einstein-Podolsky-Rosen-Bohm *Gedankenexperiment*: A New Violation of Bell's Inequalities." *Physical Review Letters*, **49**: 91–94.

Bell, J. (1987). "The Theory of Local Beables." Pp. 52–62; "How to Teach Special Relativity." Pp. 67–80 in *Speakable and Unspeakable in Quantum*. Cambridge: Cambridge University Press.

Bell, J. (1990). "Against Measurement." Pp. 17–32 in A. Miller (ed.), *Sixty-Two Years of Uncertainty*. New York: Plenum Press.

Beller, M. (1999). *Quantum Dialogues*. Chicago: University of Chicago Press.

Bohr, N. (1983). "Discussion with Einstein on Epistemological Problems in Atomic Physics." Pp. 9–49 in J. Wheeler and W. Zurek (eds.), *Quantum Theory and Measurement*. Princeton, NJ: Princeton University Press.

Bohr, N. and Rosenfeld, L. (1957). "On the Question of the Measurability of Electromagnetic Field Quantities." Pp. 357–400 in R. Cohen and J. Stachel (eds.), *The Selected Papers of Leon Rosenfeld*. Dordrecht: Reidel [1933].

Born, M. (1926). "Zur Quantenmechanik der Stossvorgänge." *Zeitschrift für Physik*, **37**: 863–867.

Bricmont, J. (1996). "Science of Chaos or Chaos in Science." Pp. 113–175 in P. Gross et al. (eds.), *The Flight from Science and Reason*. New York: The New York Academy of Science.

Bronstein, M. (1936). "Quantentheorie Schwacher Gravitationsfelder." *Physikalische Zeitschrift der Sowjetunion*, **9**: 140–157.

Brown, H. and Redhead, M. (1981). "A Critique of the Disturbance Theory of the Indeterminacy of Quantum Mechanics." *Foundations of Physics*, **11**: 1–20.

Buks, E., Schuster, E., Heiblum, M., Mahalu, D., and Umansky, V. (1998). "Dephasing in Electron Interference by a 'Which-Path' Detector." *Nature*, **391**: 871–874.

Busch, P., Heinoen, T., and Lahti, P. J. (2007). "Heisenberg's Uncertainty Principle." *Physics Reports*, **452**: 155–176.

Carazza, B. and Kragh, H. (1995). "Heisenberg's Lattice World: The 1930 Theory Sketch." *American Journal of Physics*, **63**: 595–605.

Caves, C., Fuchs, C., and Schack, R. (2002). "Quantum Probabilities as Bayesian Probabilities." *Physical Review A*, **65**: 022305.

Darrigol, O. (1992). *From C-numbers to Q-numbers: The Classical Analogy in the History of Quantum Theory*. Berkeley: University of California Press.

de Finetti, B. (1980). "Foresight. Its Logical Laws, Its Subjective Sources." Pp. 53–118 in H. E. Kyburg Jr. and H. E. Smokler (eds.), *Studies in Subjective Probability*. Huntington, NY: Robert E. Krieger Publishing Company [1937].

Dirac, P. (1927). "The Physical Interpretation of Quantum Dynamics." *Proceedings of the Royal Society of London A*, **118**: 621–641.

Dirac, P. (1929). "Quantum Mechanics of Many Electron Systems." *Proceedings of the Royal Society of London A*, **123**: 714–733.

Dirac, P. (1938). "Classical Theory of Radiating Electrons." *Proceedings of the Royal Society of London A*, **167**: 148–169.

Dirac, P. (1963). "The Evolution of the Physicist's Picture of Nature." *Scientific American*, **208**: 45–53.

Duncan, A. and Janssen, M. (2013). "(Never) Mind Your p's and q's: Von Neumann versus Jordan on the Foundations of Quantum Theory." *European Journal of Physics H*, **38**: 175–259.

Earman, J. and Redei, M. (1996). "Why Ergodic Theory Does not Explain the Success of Equilibrium Statistical Mechanics." *British Journal of Philosophy of Science*, **47**: 63–78.

Feynman, R. (1965). *The Character of Physical Law*. Cambridge, MA: MIT Press.

Feynman, R. and Hibbs, A. R. (1965). *Quantum Mechanics and Path Integrals*. New York: Dover Publications.

Fredkin, E. (1990). "Digital Mechanics: An Informational Process Based on Reversible Universal Cellular Automata." *Physica D*, **45**: 254–270.

Frigg, R. (2007). "Probability in Boltzmannian Statistical Mechanics." Pp. 92–118 in G. Ernst and H. Üttemann (eds.), *Time, Chance and Reduction, Philosophical Aspects of Statistical Mechanics*. Cambridge: Cambridge University Press.

Gieser, S. (2005). *The Innermost Kernel: Depth Psychology and Quantum Physics: Wolfgang Pauli's Dialogue with C.G. Jung*. Berlin: Springer.

Goldstein, S. (2012). "Typicality and Notions of Probability in Physics." Pp. 59–71 in Y. Ben-Menahem and M. Hemmo (eds.), *Probability in Physics*. Berlin: Springer.

Goldstein, S., Dürr, D., and Zanghi, N. (1992). "Quantum Equilibrium and the Origin of Absolute Uncertainty." *The Journal of Statistical Physics*, **67**: 843–907.

Hacking, I. (1975). *The Emergence of Probability*. Cambridge: Cambridge University Press.

Hagar, A. (2003). "A Philosopher Looks at Quantum Information Theory." *Philosophy of Science*, **70**: 752–775.

Hagar, A. (2008). "Length Matters: The Einstein-Swann Correspondence." *Studies in the History and Philosophy of Modern Physics*, **39**: 532–556.

Hagar, A. (2014). *The Quest for the Smallest Length in Modern Physics*. Cambridge: Cambridge University Press.

Hagar, A. and Hemmo, M. (2006). "Explaining the Unobserved. Why Quantum Theory Ain't Only about Information." *Foundations of Physics*, **36**: 1295–1324.

Hagar, A. and Hemmo, M. (2013). "The Primacy of Geometry." *Studies in the History and Philosophy of Modern Physics*, **44**: 357–364.

Hagar, A. and Sergioli, G. (2014). "Counting Steps: A New Interpretation of Objective Chance in Statistical Physics." *Epistemologia*, **37**: 262–275. See appendix in arXiv:1101.3521.

Heisenberg, W. (1930). *The Physical Principles of Quantum Mechanics*. New York: Dover Publications.

Heisenberg, W. (1983). "The Physical Content of Quantum Kinematics and Mechanics." Pp. 62–84 in J. Wheeler and W. Zurek (eds.), *Quantum Theory and Measurement*. Princeton, NJ: Princeton University Press [1927].

Hemmo, M. and Shenker, O. (2012). *The Road to Maxwell's Demon*. Cambridge: Cambridge University Press.

Hilgevoord, J. and Uffink, J. (1983). "Overall Width, Mean Peak Width, and the Uncertainty Principle." *Physics Letters A*, **95**: 474–476.

Hilgevoord, J. and Uffink, J. (1988a). "Interference and Distinguishability in Quantum Mechanics." *Physica B*, **151**: 309–313.

Hilgevoord, J. and Uffink, J. (1988b). "The Mathematical Expression of the Uncertainty Principle." Pp. 91–114 in A. van der Merwe, F. Selleri, and G. Tarozzi, (eds.), *Microphysical Reality and Quantum Description*. Dordrecht: Kluwer.

Hilgevoord, J. and Uffink, J. (1989). "Spacetime Symmetries and the Uncertainty Principle." *Nuclear Physics B (Proc. Sup.)*, **6**: 246–248.

Hilgevoord, J. and Uffink, J. (1990). "A New View on the Uncertainty Principle." Pp. 121–139 in A. E. Miller (ed.), *Sixty-Two Years of Uncertainty, Historical and Physical Inquiries into the Foundations of Quantum Mechanics*. New York: Plenum Press.

Ismael, J. (2009). "Probability in Deterministic Physics." *Journal of Philosophy*, **106**: 89–108.

Jammer, M. (1974). *The Philosophy of Quantum Mechanics*. New York: John Wiley & Sons.

Kennard, E. H. (1927). "Zur Quantenmechanik Einfacher Bewegungstypen." *Zeitschrift für Physik*, **44**: 326–352.

Kochen, S. and Specker, E. (1967). "The Problem of Hidden Variables in Quantum Mechanics." *Journal of Mathematics and Mechanics*, **17**: 59–87.

Kragh, H. (1995). "The Search for a Smallest Length." *Revue d'Histoire des Sciences (Paris)*, **48**: 401–434.

Landau, L. and Peierls, R. (1983). "Extensions of the Uncertainty Principle to Relativistic Quantum Theory." Pp. 465–476 in J. Wheeler and W. Zurek (eds.), *Quantum Theory and Measurement*. Princeton, NJ: Princeton University Press [1931].

Laplace, P. S. (1902). *A Philosophical Essay on Probabilities*. New York: John Wiley [1814].

Lewis, D. (1986). *Philosophical Papers Vol. 2*. Oxford: Oxford University Press.

Loewer. B. (2001). "Determinism and Chance." *Studies in History and Philosophy of Modern Physics*, **32**: 609–620.

March, A. (1951). *Quantum Mechanics of Particles and Wave Fields*. New York: John Wiley.

Maudlin, T. (2007). "What Could Be Objective about Probabilities?" *Studies in the History and Philosophy of Modern Physics*, **38**: 275–291.

Mead, C. (1964). "Possible Connection between Gravitation and Fundamental Length." *Physical Review*, **135**: 849–862.

Messiah, A. (1961). *Quantum Mechanics Volume II*. New York: Interscience Publishers.

Pauli, W. (1979). In K. Meyenn (ed.), *Wissenschaftlicher Briefwechsel mit Bohr, Einstein, Heisenberg. Band I: 1919–1929*. Berlin: Springer Verlag.

Pauli, W. (1985). In K. Meyenn (ed.), *Wissenschaftlicher Briefwechsel mit Bohr, Einstein, Heisenberg. Band II: 1930–1939*. Berlin: Springer Verlag.

Pauli, W. (1994). *Writings on Physics and Philosophy*. Berlin: Springer.

Pitowsky, I. (1985). "On the Status of Statistical Inferences." *Synthese*, **63**: 233–247.

Pitowsky, I. (1989). *Quantum Probability–Quantum Logic*. Berlin: Springer Verlag.

Pitowsky, I. (1994). "George Boole's 'Conditions of Possible Experience' and the Quantum Puzzle." *British Journal for the Philosophy of Science*, **45**: 95–125.

Pitowsky, I. (1996). "Laplace's Demon Consults and Oracle: The Computational Complexity of Prediction." *Studies in the History and Philosophy of Modern Physics*, **27**: 161–180.

Pitowsky, I. (2012). "Typicality and the Role of the Lebesgue Measure in Statistical Mechanics." Pp. 41–58 in Y. Ben-Menahem and M. Hemmo (eds.), *Probability in Physics*. Berlin: Springer.

Robertson, H. P. (1929). "The Uncertainty Principle." *Physical Review*, **34**: 163–164.

Ruark, A. (1928). "The Limits of Accuracy in Physical Measurements." *Proceedings of the National Academy of Sciences*, **14**: 322–328.

Schrödinger, E. (1930). "Zum Heisenbergschen Unschärfeprinzip." *Berliner Berichte*, 296–303.

Sklar, L. (1993). *Physics and Chance*. Cambridge: Cambridge University Press.

Stachel, J. (1993). "The Other Einstein." *Science in Context*, **6**: 275–290.

Toffoli, T. (1984). "Cellular Automata as an Alternative to (Rather than an Approximation of) Differential Equations." *Physica D*, **10**: 117–127.

Uffink, J. (1985). "Verification of the Uncertainty Principle in Neutron Interferometry." *Physics Letters A*, **108**: 59–62.

Uffink, J. (2011). "Subjective Probability and Statistical Physics." Pp. 25–50 in C. Beisbart and S. Hartmann (eds.), *Probabilities in Physics*. Oxford: Oxford University Press.

Uffink, J. and Hilgevoord, J. (1985). "Uncertainty Principle and Uncertainty Relations." *Foundations of Physics*, **15**: 925–944.

von Liechtenstern, C. R. (1955). "Die Beseitigung von Wlderspruchen bei der Ableitung der Unsch£rferelation." Pp. 67–70 in *Proceedings of the Second International Congress of the International Union for the Philosophy of Science (Zurich 1954)*. Neuchatel: Editions du Griffon.

von Neumann, J. (1932). *Mathematical Foundations of Quantum Theory*. Princeton, NJ: Princeton University Press.

Wataghin, G. (1930). "Über die Unbestimmtheitsrelationen der Quantentheorie." *Zeitschrift für Physik*, **65**: 285–288.

Wataghin, G. (1934a). "Bemerkung über die Selbstenergie der Elektronen." *Zeitschrift für Physik*, **88**: 92–98.

Wataghin, G. (1934b). "Über die Relativitiche Quantenelectrdynamik und die Ausstarhlung bei Stösen sher Energierecher Elektronen." *Zeitschrift für Physik*, **92**: 547–560.

Wigner, E (1979). *Symmetries and Reflections*. Woodbridge, CT: Ox Bow Press.

9

Inferential versus Dynamical Conceptions
of Physics

DAVID WALLACE

1 Introduction

Sometimes progress, especially in foundational matters, can come not from resolving a dispute, but by clarifying its structure. This chapter attempts to do this in the cases of foundations of statistical mechanics (SM) and quantum mechanics (QM). Its main theme is that we can identify two attitudes to a given area of physics – *inferentialism*, the idea that a given theory is a tool to allow us to make inferences about present and future facts or experiments, and *dynamicism*, the idea that a theory is an account of the dynamical behavior of systems entirely independent of our own knowledge.

It may sound from this as if "dynamicism" and "inferentialism" are just new names for "realism" and "instrumentalism." But the distinction here is local (i.e., relative to a given area of physics) and indeed the "inferences" being drawn are in general supposed to be inferences about some other area of physics construed realistically – so that, for instance, much discussion of classical statistical mechanics is tacitly or explicitly inferentialist even when those discussing are fully committed to realism about classical physics. I will argue that the divide between inferentialism and dynamicism illuminates the debate both in statistical and in quantum foundations. Partly this hoped-for illumination occurs within each separate subject: I will try to show that the inferential-vs.-dynamical dispute naturally captures much of the basic disagreement within statistical mechanics, and that discussions of the quantum measurement problem often presuppose one or other conception even in stating the *problem*, so that proposed *solutions* can be misunderstood. More interestingly, it illustrates the possibility of close links between strategies in one case and in the other.

I make the case for this in three parts. In Section 2, I explain how the two rival conceptions play out in classical statistical mechanics, and canvass the main problems with each. In Section 3, I do likewise for quantum mechanics, and consider by analogy how those "main problems" look in a quantum-mechanical

179

context. At this point, I hope to have established strong *similarities* between the inferential/dynamical dispute in the two fields.

However, it is possible to go beyond similarities. When we consider not classical but quantum statistical mechanics, there is an almost complete collapse of the statistical-mechanical questions onto the quantum questions, to the point that it becomes essentially impossible to adopt the inferential conception in the one case and the dynamical conception in the other. In the process of establishing this in Section 4, I will argue that while classical statistical mechanics can be considered as a probabilistic extension of classical mechanics, the same is not at all true in quantum statistical mechanics (I elaborate on this point in Wallace 2016a).

I conclude that a single interpretative question – whether to conceive of a given field in physics as a form of inference or as a study of dynamics – plays a central role in the foundations of quantum theory, and *the exact same role* in the foundations of statistical mechanics once it is understood quantum mechanically. I conclude by drawing some morals for the study of the conceptual foundations of both fields.

2 Conceptions of Statistical Mechanics

In trying to get clear on what (classical) SM is, we can identify two main themes. The first starts from the observation that in macroscopically large systems we are necessarily ignorant of many features of the system:

(1) We do not know its present micro-state.
(2) We do not know its exact Hamiltonian.
(3) Even if we did know (1) and (2), we do not know how to solve the exact equations of motion in order to predict the system's future micro-state.

SM, on this account, is concerned with how we can make correct, or at any rate reasonable, inferences about macroscopic systems in the face of these epistemic limitations. SM, that is, is a branch of the general theory of *inference* under conditions of imperfect information. To quote a classic account:

The science of SM has the special function of providing reasonable methods for treating the behaviour of mechanical systems under circumstances such that our knowledge of the condition of the system is less than the maximal knowledge which would be theoretically possible. The principles of ordinary mechanics may be regarded as allowing us to make precise predictions as to the future state of a mechanical system from a precise knowledge of its initial state. On the other hand, the principles of SM are to be regarded as permitting us to make reasonable predictions as to the future condition of a system, which may be expected to hold on the average, starting from an incomplete knowledge of its initial state.

The other concept begins from the empirical observation (going back at least to the nineteenth-century development of thermodynamics) that the collective degrees of freedom of macroscopic systems demonstrate observed, law-like regularities: ice cubes melt in water, gases expand to fill boxes, heat flows along metal bars in accordance with the diffusion equation, and so forth. On this conception of SM, its goal is to derive these collective dynamical results from the underlying micro-dynamics. In principle this might be an exceptionless derivation, much as we can derive exceptionless equations for the movement of the center of mass of a general body, or the rotational dynamics of a rigid body, even when that body is made of a very large number of components. But in practice there are strong reasons to expect any such statistical-mechanical derivation to require additional assumptions to be made.

On this account, SM is a branch of the general theory of *dynamics*: it studies a sub-problem of the general problem of understanding the behavior of systems over time. A classic statement of *this* position is the following:

[I]n the case of a gas, consisting, say, of a large number of simple classical particles, even if we were given at some initial time the positions and velocities of all the particles so that we could foresee the collisions that were about to take place, it is evident that we should be quickly lost in the complexities of our computations, if we tried to follow the results of such collisions through any extended length of time. Nevertheless, a system such as a gas composed of many molecules is actually found to exhibit perfectly definite regularities in its behaviour, which we feel must be ultimately traceable to the laws of mechanics even though the detailed application of these laws defies our powers. For the treatment of such regularities in the behaviour of complicated systems of many degrees of freedom, the methods of statistical mechanics are adequate and especially appropriate.

The two conceptions are not always cleanly separated – indeed, my two quotes come from consecutive pages of the same classic textbook (Tolman 1938: 1–2) – and it may be that a successful account of statistical mechanics can be given that combines aspects of both. But prima facie they are rivals, not complements, and in these cases (and, I think, many others) the appeal to aspects of both seems to involve equivocation and confusion rather than clear-eyed synthesis. Evidence of their prima facie incompatibility comes from their very different foundational implications, as a few examples should make clear.

2.1 *The Conceptual Status of Probability*

On the inferential conception, we are assumed to have imperfect knowledge of at least the current micro-state. It is therefore natural mathematically to introduce probability measures to represent this imperfect knowledge: the probability assigned to a given micro-state represents our level of confidence that the system is really in that micro-state. To borrow a very useful piece of terminology from the quantum foundations literature: we can distinguish the *ontic state* of the system, which represents what properties and features the system actually has, from the *epistemic state*, which represents our ignorance of the ontic state. The ontic state is given by a phase-space *point*, the epistemic state by a phase-space *probability distribution*. (There is then a secondary question about how constrained our choice of epistemic state is: on the influential "objective" approach championed by Edward Jaynes 1957a, 1957b, and the papers in 1983, it is specified uniquely, for given information, by a combination of maximum-entropy principles and considerations of symmetry; on a more "subjective" approach, two agents with the same information might rationally disagree. I will be concerned here in particular with Jaynes's version of the inferential conception, which has been hugely influential in the physics literature.)

On the dynamical conception, there is no up-front need for probabilities. We are interested in the large-scale features of the system, and this may require *statistical* considerations – what fraction of molecules in a gas have a given velocity, for instance – but these can be understood as categorical features of the system. Boltzmann's original discussion of the *H* theorem is generally thought to have had this form, in particular (see Brown, Myrvold, and Uffink 2009 and references therein for historical discussion): Boltzmann took himself to be deriving, under reasonable assumptions, the conclusion that the statistical distribution of molecular velocities in a dilute gas would *reliably* approach the Maxwell-Boltzmann distribution.

However, it is well known that Zermelo and Loschmidt's objections established that Boltzmann's results could not be exceptionless facts about the dynamics of dilute gases, but could only hold given certain other assumptions – and Boltzmann's and later attempts to fill in these assumptions have invariably turned out to be probabilistic, at least in part. That is, the dynamical conception ends up making claims not about how a system will *invariably* behave, but about how it will *most probably* behave. Furthermore, while *this* requirement for probability is a requirement for a probabilistic micro-foundation for a deterministic macro-prediction (the approach to equilibrium of dilute gases), plenty of the applications of SM (notably, fluctuation phenomena) also lead to probabilistic predictions (for more on this point, see Wallace 2015). So while the dynamical approach does not require probability a priori, it has in fact proven to be an essential component of SM. Indeed, the machinery of contemporary SM makes very extensive use of probability measures over phase space: any attempt to make sense of that machinery must make sense of those probabilities.

But on the dynamical conception, how are these probabilities to be understood? Not as representations of our ignorance of the true micro-state (on pain of collapsing the dynamical conception into the inferential one). For some time, a popular suggestion was that they should be understood as long-time averages, but this is beset with conceptual and technical problems, notably in the interpretation of non-equilibrium SM. The language of "ensembles" suggests that they are to be understood as relative frequencies in a fictional, infinite collection of copies of the system – but if the collection is fictional, how are we to understand its relation to the single, actual system? If there is anything physical in the world corresponding to these probabilities, the only obvious candidate is relative frequencies in actual systems – but there are well-known problems (see, e.g., Hajek 1996, 2009) in identifying probability with frequency in this way.

This is perhaps a good point to note that the division between inferential and dynamical conceptions of SM was famously portrayed by Jaynes (1965) as a division between Gibbsian and Boltzmannian versions of SM, and that this portrayal remains common in the literature (which is not to deny that the difference between the two approaches can be discussed in an interpretation-independent way). Consider, for instance, Goldstein (2001), in a section titled "Boltzmann's Entropy vs. the Gibbs Entropy":

It is widely believed that thermodynamic entropy is a reflection of our ignorance of the precise microscopic state of a macroscopic system, and that if we somehow knew the exact phase point for the system, its entropy would be zero or meaningless. But entropy is a quantity playing a precise role in a formalism governing an aspect of the behavior of macroscopic systems. This behavior is completely determined by the evolution of the detailed microscopic state of these systems, regardless of what any person or any other being happens to know about that state.

(See also Albert 2000: 67–68.) This has the disadvantage of conflating the *use* of certain bits of mathematical machinery with the *interpretation* placed on that machinery, suggesting that a defender of the dynamical conception of SM has to reject use of probability distributions, and indeed that they have to reject or reconstruct a large fraction of contemporary results in SM. In Wallace (2016b) I argue that Gibbsian and Boltzmannian mathematical methods can both be understood within a dynamical perspective, and indeed that from that perspective the differences between them are relatively minor; see also Lavis (2005) for another (somewhat different) strategy for the reconciliation of the two approaches.

2.2 Equilibrium

The foundational assumption of thermodynamics, inherited by equilibrium SM, is that isolated systems can in general, and perhaps after a certain period of time has elapsed, be assumed to be in *equilibrium*: to be in a state whose macroscopic parameters are unchanging with time. SM offers a concrete characterization of this equilibrium in terms of a probability distribution – the *microcanonical ensemble*, in the simplest case. It is a well-confirmed empirical fact that measurements of the values of macroscopic parameters of equilibrium systems, and of the fluctuations in those values, are accurately predicted by the microcanonical distribution (the *Boltzmannian* characterization of equilibrium defines a system as at equilibrium not when its probability distribution is microcanonical, but when its micro-state is located in a certain phase-space region. But even Boltzmannians need to recover fluctuation phenomena, and so must require that an isolated system left to its own devices for the equilibration timescale has a probability of being in a given phase-space region given by the microcanonical distribution – see Wallace 2016b). Both inferentialists and dynamicists must account for these facts. They do so, however, in very different ways.

For inferentialists, that a system is at equilibrium is a largely a priori matter. According to the maximum-entropy principle defended by Jaynes (and finessing certain measure-theoretic concerns), the equilibrium probability distribution is the rationally required distribution that represents my ignorance of the micro-state of a system when all I know of that system is its energy. To say "system X is at equilibrium," for the (Jaynesian) inferentialist, is to say just "all I know of system X is its energy." As such, the question of how systems approach, or end up at, equilibrium is thus not obviously well-posed on the inferentialist conception: whether a system is at equilibrium or not is a property not of the system alone, but of the observer's knowledge of the system.

For dynamicists, on the other hand, whether or not a system is at equilibrium is a contingent claim about that system alone, and the claim that isolated systems should in general be treated as being at equilibrium is justified only if it can be established (on whatever additional assumptions are required) that isolated systems initially out of equilibrium approach equilibrium in a reasonable length of time (at least with high objective probability, however that is to be understood). The microcanonical characterization of equilibrium is legitimate only if it is the unique probability distribution to which other probability distributions evolve in isolated systems (perhaps at some coarse-grained level of approximation rather than exactly). (Similarly, the Boltzmannian claim that equilibrium is characterized by the largest macrostate is legitimate only if it is dynamically the case, on reasonable assumptions, that an isolated system's micro-state is highly likely to evolve into that macrostate.) The foundations of equilibrium, on the dynamical conception of SM, thus require (in principle) quite detailed considerations of a system's dynamics: the Boltzmann equation can be seen as an early prototype of these kind of considerations, and the long-running explorations of ergodicity and mixing (see Sklar 1993: ch. 7) can be seen to be more modern attempts, however well or badly motivated those particular strategies might be.

2.3　*Thermodynamic Entropy and the Second Law*

On the inferential conception, the Gibbs entropy,

$$S_G(\rho) = -k_B \int dx\, \rho(x)\, \ln\rho(x), \tag{1}$$

is commonly identified with the thermodynamic entropy (albeit Gibbs himself did not so identify them; see Uffink 2007: section 5.1). This entropy, being a functional of the probability distribution, represents (on the inferential conception) a measure of an agent's lack of information about the underlying micro-state: the more widespread and uniform the probability distribution, the higher the entropy. This interpretation of entropy as a measure of negative information can be formalized: on certain assumptions (notably including a designation of the Liouville measure as the preferred measure of uniformity on phase space) and up to a constant factor, it can be shown to be the unique such measure (Jaynes 1957a). It is maximized, under the constraint that the energy has a given fixed value, by the microcanonical distribution, and this provides the justification for the inferentialist of using that distribution to represent equilibrium.

The Gibbs entropy is well known to be invariant under Hamiltonian time evolution. It follows that if a system initially at equilibrium at energy E and external parameters (say, volume) V (i.e., a system about which the observer knows only the energy and the volume) is allowed to evolve under Hamiltonian flow (perhaps involving external potentials) until it has energy E' and volume V', then the Gibbs entropy of the resultant distribution is less than or equal to the Gibbs entropy of the equilibrium distribution defined by E' and V'. It has been argued (Jaynes 1957a) that this provides a justification for the Second Law of Thermodynamics on the inferential conception. The obvious problem with this account is that while the new equilibrium distribution has higher entropy than the original equilibrium distribution, the actual new distribution – which represents the observer's information about the current state given that at an earlier time it had energy E and volume V – has the same entropy as the original distribution. The increase in entropy seems to occur because the observer discards some information (that the system came to have energy E' and volume V' in a specific way), rather than because of any feature of the system itself. Possible solutions to this problem include appealing to our lack of knowledge of the exact dynamics (Peres 1993: 353–354) or to our lack of ability to perform the actual calculations that generate the time-evolved distribution or even to store the information about it (Myrvold 2016).

The Gibbs entropy does not seem suitable to represent thermodynamic entropy on the dynamical conception: there the probabilities must be understood (somehow) as objective features of the system, and so the Gibbs entropy is a constant of the motion. There are basically two resolutions of this problem. First, we can define a coarse-graining map or Zwanzig projection (see Zwanzig 1961) J that smooths out the fine details of a probability distribution ρ, and define the coarse-grained Gibbs entropy

$$S_{G;J}(\rho) = -k_B \int dx \; (J\rho)(x) \; \ln(J\rho)(x). \tag{2}$$

It is important to note that this is still a functional of ρ: J is to be understood as a mathematical operation used in the definition of the entropy, not as a literal transformation of the underlying probabilities. (Note that on the *inferential* conception this latter interpretation of J might be acceptable – representing, say, our finite powers of resolution of the system's detail – but it is incompatible with the *dynamical* conception as long as the underlying dynamics are taken to be Hamiltonian.)

The second resolution rejects the Gibbs entropy entirely, and instead adopts Boltzmann's old definition: phase space is divided into cells ("macrostates") and the Boltzmann entropy $S_B(M)$ of a macrostate is $k_B \times$ the logarithm of its phase-space volume. The Boltzmann entropy $S_B(x)$ of a micro-state x is then just the Boltzmann entropy of the unique macrostate in which x lies: this definition of entropy is entirely non-probabilistic. (For this reason, it is often argued – Albert 2000; Callender 2001; Goldstein 2001; Lebowitz 2007 – that the Boltzmann definition is preferable, and indeed that the Gibbsian definition only makes sense on an inferentialist conception of SM; I criticize this view in Wallace 2016b.)

In both cases, there is a considerable technical project left to carry out. First, the correct notion of coarse-graining or macrostate partitioning must be found. Often the criteria for this notion are stated in epistemic terms (states are in the same macrostate if they are "macroscopically indistinguishable" or some such), but this is an unnecessary concession to inferentialism: on the dynamicist conception, the criterion for a coarse-graining being correct (as in any case of emergence) is simply that we can write down robust dynamical equations for the collective degrees of freedom that abstract away from irrelevant micro-level details.

Second, we have to show, or at least make plausible, that entropy thus defined does indeed increase under the transformations typical of thermodynamics. Neither $S_{G;J}$ nor S_B is a constant of the motion, so there is no a priori barrier to showing that they increase in such circumstances, but to show that they in fact *do* increase requires engagement with the dynamical details of the system (for all that it can be made extremely plausible in many cases).

The significantly larger technical burden of accounting for the Second Law on the dynamicist conception might be taken to be a strength or a weakness as compared to inferentialism: a weakness, if you regard it as undesirable to have to get tangled up in the messy details of the dynamics; a strength, if you believe that the validity of thermodynamics must ultimately lie in the dynamics and that attempts to bypass the messy details have "the advantages of theft over honest toil" (Russell 1919: 71).

2.4 Retrodiction and Time Asymmetry

The underlying micro-physics, under either conception of (classical) SM, makes no particular distinction between past and future; but the actual universe shows, and SM models, manifestly time-asymmetric processes. The clearest example here is the approach to equilibrium: systems not currently at equilibrium generally evolve to equilibrium (and we have seen how each conception of SM attempts to explain this); why, by parity of reasoning, should we not expect that systems currently not at equilibrium were at equilibrium in the past? More generally, insofar as SM underpins non-equilibrium dynamical processes like the expansion of gases into an empty space, why does that reasoning not likewise work into the past?

On the inferential conception, the paradox might be put as follows: let's stipulate that given our information about the present-day state of a system, it is reasonable to infer that we should have a high degree of belief that the system is in such-and-such state in the future. For instance, suppose that our current information about a glass of water is that it contains an ice cube and some warm water (of given volumes and average energies); let us stipulate that it is reasonable on that information to infer that time $+t$ from now the ice cube is melted and the water is cooler. Why is it not equally reasonable to infer the same facts at time $-t$?

There is, on the face of it, a fairly straightforward answer. In fact, in realistic situations we have a great deal of information about the glass of water over and above its present state: we probably know what its state was five minutes ago, for instance, and even if we do not, we know a great deal of other facts about the state of the world five minutes ago. It turns out (and this is much of what gives SM its power) that this additional information about the past is of negligible importance in making predictions about the glass's future state – but it is crucially important in making predictions about its past state. The asymmetry in our *inferences* is caused by an asymmetry in our *information*.

We can question whether we really have information about the past. After all, we have no *direct* access to the past, but only to our memories of it, and those are presumably coded in the present state of the world (in particular, in our brains and our external records). And if all information is ultimately present-day information, the "asymmetry of information" explanation for the asymmetry of inference seems to be in trouble.

Here's another way to put the same problem, which makes more stark its paradoxical aspects. There are good reasons to think that, conditional on the present-day macroscopic facts about the world and on the uniform (Liouville) probability measure over micro-states consistent with that macrostate, it is overwhelmingly more likely for our records of apparent past events to have spontaneously fluctuated into existence than it is for them to be consequences of those apparent events. So what justifies our belief that the past really happened?

Put this way, we seem to have a problem in epistemology rather than one in physics: the position that the past did not happen, after all, seems to be a form of skepticism, and the question of whether we are justified in treating our memories as information directly about the past or as information about our present-day brain state is a question about how to set up our epistemology. But something strange has happened here. For the facts that ice melts in water, that stars radiate light rather than absorbing it, that people grow old, and so forth, certainly appear to be facts about the world rather than about our means of inferring things about the world. So any explanation of them that relies for a success on a certain analysis of our epistemology seems problematic. Furthermore, as a matter of logic no such explanation can explain *why* we as agents and as observers are in an asymmetric epistemic position. If observers, too, are just physical systems, we ought to explain why the system as a whole (including, inter alia, both the glass of water and the physicist contemplating it) displays the asymmetry in time that it does.

It's tempting at this point just to shrug and say that such an ineliminable role for the observer was always part and parcel of inferentialism. But this is too quick: the "inferential conception" is here a conception of SM, not of physics as a whole. Recall the distinction between ontic and epistemic states: the real, objective physics concerns the ontic state and its dynamics, and is represented by classical mechanics. SM is simply an inferential layer applied to that underlying reality to address our imperfect information about it. The problem for the inferentialists is not that their program makes use of notions like "observer" and "equilibrium": it is that various facts about the world – and, in particular, its apparent time asymmetry, even in situations where no human intervention is occurring – seem to belong more naturally to the observer-independent, ontic part of the theory, not to the epistemic part.

What of the dynamical conception? Here the paradox is more direct, because the question of whether we can derive a given time-asymmetric result is a question about objective facts about the world. If, for instance, at time 0 we can derive, under certain time-symmetric and time-translation-invariant assumptions, that entropy $S(t)$ (either Boltzmannian or coarse-grained-Gibbsian) satisfies $S(+t) > S(0)$ for time $t > 0$, the symmetry of the underlying dynamics tells us that $S(-t) > S(0)$ for $t > 0$ also. And then by time *translation* symmetry $S(0) > S(+t)$, and we have no paradox, but a straightforward algebraic contradiction. So our derivation must build in some violation of time-reversal or time-translation invariance and it is simply a matter of finding it.

This task is, in fact, straightforward. The dynamicist approach to probability requires an assumption that the probability distribution at the initial time has certain features (such as being uniform across a given set of micro-states), and these features are not in general preserved under time evolution. If we impose this required assumption at a given time, it will deliver the required dynamical results at later times but not at earlier times. So insofar as the assumption is to be thought of as a *physically contentful* boundary condition, and not simply a statement about our information, it must follow that we have to impose that boundary condition at the earliest relevant time for the system under study. And since boundary conditions cannot be set for each system separately in an interconnected universe, ultimately it seems that the dynamical conception requires a particular condition to be imposed on the initial state of the Universe. (Mathematically speaking the condition could be seen instead as being imposed at the time at which a system is prepared (this is how Malament and Zabell 1980 and Vranas 1998 state their probability conditions), but this introduction of an explicit process of "system preparation" imposed externally and primitively seems to fall under the inferentialist rather than the dynamicist framework.)

The most common choice of the condition on the initial state of the Universe is the dual requirement that (i) the initial macrostate of the Universe was a particular low-Boltzmann-entropy state (the "Past Hypothesis"), and (ii) the initial probability distribution is uniform or at least reasonably smooth over that macrostate (in Wallace 2010b I argue that (i) is not in fact needed). In practice defenses of the Past Hypothesis often seem to equivocate between dynamical and inferential conceptions of SM: often the probability measure seems to be argued for a priori and the Past Hypothesis is then justified on epistemic grounds, to avoid global skepticism.

2.5 *Compare and Contrast?*

So: in the cases of probability, of equilibrium, of entropy, and of time asymmetry – that is, in the main loci of discussion in foundations of SM – the inferential and dynamical conceptions of SM give very different accounts of what is going on. Which conception works best? So far as I can see, the most severe problems with each are as follows:

- The most serious problem for the dynamical conception is *probability*. SM is suffused with probabilistic claims and couched in probabilistic language. Yet it is extremely difficult to see just how these probabilities are to be understood, if not as some quantification of our imperfect information about the actual state. And what role can a probability distribution over such states play in an explanation of the actual dynamical behavior of the system, given that how it evolves in the future depends entirely on its actual, unique state?

(Note that this is not simply the general philosophical problem of how objective chance can be understood in science. It is one thing to say of a system whose present state is x that it has a certain objective probability of transitioning in the next instant to a state x'. It is quite another to say of that system that it has a certain objective probability of currently being in state x'.)

Advocates of a dynamical conception are sensitive to this worry – it drives part of the widespread skepticism about Gibbs entropy and the corresponding preference for Boltzmann entropy – but simply defining entropy that way will not, by itself, suffice to remove the reliance on probability in SM.

- The most serious problem for the inferential conception is *objectivity*, and the most dramatic and serious example of the problem is asymmetry in time. To the inferentialist, virtually none of the claims made in SM are claims about the world in itself, but just about how I should reason about the world given imperfect information. This already has a somewhat problematic feel in the inferentialist *characterization* of equilibrium not as a state that systems generally speaking *in fact* get into and remain in, but just as a way of saying that we don't know anything about a system's state. It comes close to paradox when we ask for an account of why the *non-equilibrium* processes in the world display a clear and consistent time asymmetry, and in particular if we ask for an account of that time asymmetry that does not make essential reference to an external, already assumed, asymmetric observer.

This should not be surprising. Classical SM seems to be a hybrid, displaying some features that suggest an inferential conception and some a dynamical one. Probability, in the classical deterministic context, is extremely difficult to understand dynamically. The time-asymmetric dynamics of the Boltzmann equation and its many relatives is extremely difficult to understand inferentially. (Equilibrium thermodynamics could be called either way: is it the study of which processes are *physically possible*, or which transformations are *within the power of an agent*.)

If classical mechanics were correct (and if, *per impossibile*, I could still exist under that assumption), I would end here with the suggestion that the inferentialist-vs.-dynamicist way of understanding the debates in SM is more helpful, and less prone to mutual miscommunication, than the Gibbs-vs.-Boltzmann approach currently prevalent. However, the move from classical to *quantum* SM radically changes the terms of the debate, as we will see. First, though, it is necessary to consider quantum theory *itself* from inferential and dynamical perspectives.

3 Conceptions of Quantum Mechanics

The basic dynamical axioms of QM are simple enough to state, and can be done in direct parallel with those of classical mechanics. Instead of a state space of phase-space points, we have Hilbert-space rays. Instead of evolution under Hamilton's equations, we have evolution under the Schrödinger equation. And to find the state space of a composite system, instead of taking the Cartesian product of phase spaces we take the tensor product of Hilbert spaces.

There is, however, a different parallel we could have drawn. Quantum states could have been considered to be analogs of *probability distributions* over phase space, not of phase space points; the Schrödinger equation could have been compared to the Liouville equation, not Hamilton's equations; the tensor-product rule for constructing the state spaces of composite systems could have been regarded as correct in QM *just as* in classical probabilistic mechanics.

At the root of the difference between the two approaches is this question: is the quantum state something like a physical state of a system, or something like a probability distribution? One way to state the notorious measurement problem (a somewhat heterodox way, to be sure; see Wallace 2012a) is that "orthodox" QM – that is, QM as it is in practice used – systematically equivocates between the two ways of understanding the state:

The state as probability distribution: The quantum state of macroscopic systems, and the quantum state of even microscopic systems in contexts of measurement, is treated as representing a probability distribution over possessed values: if a system is a macroscopic superposition of, say, "cat alive" and "cat dead" states, we treat the cat as *either alive or dead*, with the probability of each equal to the mod-squared amplitude of the corresponding state in the superposition. Similarly, we analyze contexts of state *preparation* as something very akin to probabilistic conditionalization: if we generate a collimated beam of atoms by putting a narrow slit in front of a furnace, we conditionalize on the fact that the particles went through the slit, and discard the part of the quantum state corresponding to the atoms *not* going through the slit. (I am grateful to Simon Saunders – in conversation – for pressing the role of state preparation here.)

The state as representing something physical: The quantum state of microscopic systems, or indeed any systems, in situations where interference phenomena occur, is (prima facie) treated as representing a physical, albeit in general highly non-classical, state of the world. In an interference experiment, we talk freely about interference between the parts of the quantum state corresponding to the various terms in the superposition, in a way that does not straightforwardly lend itself to probabilistic reinterpretation.

In this way of looking at things, the "measurement problem" is simply this incoherence about how we are to understand the quantum state: what are the criteria for when we should treat it as a probability distribution and when as a physical state, and how can we make sense of the transition between the two conceptions?

In the space of moves made to resolve/solve/dissolve the measurement problem, I believe we can identify two broad themes, which roughly track the two themes discussed in statistical mechanics; I discuss each in turn.

3.1 The Inferential Conception of QM

On this conception, the quantum state does not represent anything physical: it is to be understood as a probability measure, and that probability measure represents in some way our degrees of belief in some set of physical facts. On this conception, macroscopic superpositions like "Schrödinger cat" states are utterly unproblematic; I represent a cat with a Schrödinger cat state just in case I am unsure as to whether it is alive or dead following some measurement process. Asher Peres expresses this position with great clarity:

[T]he "cat paradox" arises because of the naive assumption that the time evolution of the state vector ψ represents a physical process which is actually happening in the real world. In fact, there is no evidence whatsoever that every physical system has at every instant a well-defined state ψ (or a density matrix ρ) and that the time dependence of $\psi(t)$ (or of $\rho(t)$) represent[s] the actual evolution of a physical process. In a strict interpretation of quantum theory, these mathematical symbols represent different *statistical information* enabling us to compute the *probabilities* of occurrence of specific events.

(1993: 373–374)

Peres also makes clear why I call this approach "inferentialist": it interprets QM not as a description of physical reality, but as a calculus to make (probabilistic) predictions in the absence of such a description. Chris Fuchs says it even more explicitly:

Quantum states are states of information, knowledge, belief, pragmatic gambling commitments, not states of nature.

(2002: section 6)

If inferentialism dissolves the problem of macroscopic superpositions, problems arise when we apply it to *microscopic* superpositions. We can in practice treat a Schrödinger cat state as a probabilistic mixture of live and dead cat without contradicting ourselves; if we try to treat an electron in a superposition of spin states as a probabilistic mixture of those states, we will get the wrong answers in our calculations. So if the probabilities are *not* probabilities for the electron to have various values of spin, what *are* they probabilities for?

By analogy with (inferentialist) SM, we can identify one apparently highly attractive possibility. In SM, we distinguished between the ontic state (representing the actual world) and the epistemic state (representing our information about, or degrees of belief about, the ontic state). It is tempting, then, to read the quantum state likewise as an epistemic state (indeed, what I call the inferentialist conception is often called the ψ-epistemic approach, as contrasted with the ψ-ontic approach, which takes the quantum state ψ as representing something physically real).

This suggests a research program: find that theory that is to QM as classical micro-dynamics is to (inferentialist) SM. This is the program that Einstein, for instance, plausibly had in mind when he spoke of "hidden variables" (see Harrigan and Spekkens 2010). But the progress made in that program, though substantial, has been largely negative, notably:

- Bell's theorem requires that the theory involve instantaneous action at a distance.
- The Kochen-Specker theorem requires that the theory have the apparently pathological feature of *non-contextuality*.
- The recently proved Pusey-Barrett-Rudolph theorem (Pusey, Barrett, and Rudolph 2011; see Maroney 2012 for discussion and development) rules out such theories entirely, given apparently very weak assumptions about our ability to treat systems as independent from one another. (The most significant positive result of which I am aware is Rob Spekkens' "toy model," 2007.)

The main alternative proposed is that the probability distribution is over possible outcomes of measurements, represented either by projection-valued measures on Hilbert Space (PVMs), or (more commonly in recent work) by positive operator valued measures (POVMs). There is no mathematical obstacle to this move; indeed, it can be proven that any such probability measure, provided that it is non-contextual, is represented by some pure or mixed quantum state (the proof requires great labor in the case of PVMs – Gleason 1957 – and is pretty straightforward for POVMS – Caves et al. 2004 – the non-contextuality requirement is admittedly difficult to justify on this conception of the state, as I discuss further in Wallace 2012b: 225–226).

The problem with this strategy is that it is hard to understand it as anything other than straight instrumentalism. Some of its advocates (e.g., Peres 1993) are happy to accept this; others (notably Fuchs 2002) reject it and maintain that their proposal is compatible with some kind of realism, but it is obscure to me exactly what they hope to be realist about (see Timpson 2008 for further discussion).

3.2 The Dynamical Conception of QM

On this conception, QM – like classical mechanics – is a dynamical theory, concerned with how the physical features of the world evolve over time without reference to any observer, and the quantum state is taken to be entirely objective. With the exception of an interpretation of the quantum state as probabilistic in an objective sense, as in the dynamicist conception of classical statistical-mechanical probabilities – this option is not much explored; the ensemble interpretation of Ballentine (1970) is the only counterexample of which I'm aware – this more or less commits advocates to accepting (i) that the quantum state is not *fundamentally* a probabilistic entity (though this is compatible with the dynamical laws *governing* the quantum state being fundamentally probabilistic, as in dynamical collapse theories) and (ii) that at least in the case of microscopic systems, it needs to be understood as physically representing their properties in the same basic way as does the classical micro-state.

The burning question for this strategy is then: how come the quantum state *appears* to be probabilistic when macroscopic systems are concerned? Here a large amount of technical progress has been achieved, under the general label of *decoherence theory*. Two broad strategies have been applied here:

- The *decoherent histories* (or consistent histories) program (see, e.g., Griffiths 1984, 1996, 2002; Omnés 1988, 1992, 1994; Gell-Mann and Hartle 1993; Halliwell 1998, 2010; Hartle 2010) has asked directly: under what conditions do the probabilities assigned by QM to sequences of observables obey the probability calculus? The general conclusion is that, at least a sufficient condition, QM delivers probabilistic (or very nearly probabilistic) predictions for the evolution of coarse-grained properties of systems: that is, it is probabilistic on the macro scale once the microscopic details are averaged over.

- The *environment-induced decoherence* program (see, e.g., Zurek 1991, 2003; Zeh 1993; Joos et al. 2003; Schlosshauer 2007) has considered how the evolution of a macroscopic-scale system is affected by its coupling to an (internal or external) environment. The main focus of the program has been the selection by the environment of a preferred basis (usually a wave-packet basis) for the system with respect to which interference is strongly suppressed. This in turn has the consequence that the system can be consistently treated as probabilistically evolving (this point is stressed by Zurek 1998; see also Wallace 2012b: ch. 3 for discussion).

As a result, we now have a reasonably good understanding of the dynamical processes by which the quantum state comes to have the structure of a probability measure with respect to macroscopic degrees of freedom, even while being non-probabilistic at the micro-level. In the process, we also learn why the in-principle-incoherent shifting between probabilistic and non-probabilistic readings of the quantum state does not lead to practical problems (see Wallace 2012a for further discussion of this point).

(Philosophers have been almost as concerned with a separate problem: if the quantum state is a physical state, how are the physical features of that state to be understood? For discussion of this point, see, e.g., Hawthorne 2010; Maudlin 2010;

Wallace 2010a, and the papers in Ney and Albert 2013. With rare exceptions, physicists have been fairly unconcerned).

It is at best controversial whether all this provides a *conceptually* satisfactory understanding of probability in QM, and of QM more generally. If the quantum state ultimately is taken to represent something physical, then the terms in a superposition each seem to represent physical features of the world. In a Schrödinger cat state, then, the fact that the amplitudes of the live-cat and dead-cat terms can be consistently treated as probabilities does not seem to conceptually justify assuming that only one represents anything physically real. That is, the dynamical conception of QM – at least as long as QM itself is unmodified at the formal level – is tantamount to the Everett interpretation. In this context, the live-cat and dead-cat terms represent physically coexistent goings-on – "branches" in the usual terminology – and it has been extensively discussed whether more than decoherence is needed to justify interpreting the branches' weights probabilistically, and if so, whether and how the gap can be filled (for an introduction, see Greaves 2007; for extensive discussion, see Saunders et al. 2010; I give my own account in Wallace 2012b: chs. 4–6).

The main alternative to the Everett interpretation within the dynamical conception is to modify quantum theory – either by adding a dynamical state-vector collapse rule to eliminate all but one term in a macroscopic superposition, or by adding hidden variables as alternative loci of probability in such a way as to give probability distributions over macroscopic states of affairs equivalent to those of some particular branch (notice that unlike the (ill-fated) inferentialist hidden-variable strategy, this version maintains the quantum state as physically real and adds additional hidden variables). In dynamical collapse theories, of which the best known are the Ghirardi-Rimini-Weber theory (Ghirardi, Rimini, and Weber 1986) and the Continuous State Localization theory (Pearle 1989), probabilities now become part of the laws of physics, via an irreducibly stochastic dynamics. In hidden-variable theories (of which the de Broglie-Bohm theory is much the best known), the probabilities normally enter via a measure over hidden variables whose interpretation recapitulates the puzzles of probability in classical SM (at least, they do so in the case of the de Broglie-Bohm theory; as discussed in Bub 1997, it is perfectly possible to construct stochastic hidden-variable theories where the probabilities have basically the same status as in dynamical-collapse theories). In either case the theory must be constructed so that the newly added probabilities are numerically equal to the branch weights given at the coarse-grained level by the quantum state.

Philosophers have generally been more concerned than physicists about the conceptual difficulties of probability in the Everett interpretation, and less concerned about the technical difficulties incurred (especially in the relativistic case) by modifications of quantum theory. But both Everett and the modificatory strategy are examples of the dynamical conception: on both, the quantum state is resolutely physical; on both, physics is concerned with the actual dynamical behavior of the world, independent of our knowledge of it (this is perhaps somewhat controversial for deterministic hidden-variable theories, where there is space for an inferential/ dynamical dispute about the nature of the probability measure over hidden variables).

3.3 Compare and Contrast? The Quantum Case

In my discussion of SM, I identified the most serious conceptual problems for the inferential and dynamical strategies as being, respectively, concerns about the objectivity of statistical mechanics' deliverances (in particular, the direction of time) and about the nature of statistical-mechanical probabilities, if they are not to be understood epistemically. How do their quantum analogs fare?

It seems to me hard to deny that objectivity is a worse problem for quantum-mechanical than for statistical-mechanical inferentialists. In the statistical-mechanical case it was clear what part of physics was objective (classical mechanics) and what part was a matter of inference about the objective part (statistical mechanics); the worry was that certain features of SM seemed to belong more naturally in the objective part. But it is extremely difficult, in the light of the various no-go theorems, to see what stands to classical mechanics as QM stands to SM.

Put another way, QM is our current general dynamical framework, replacing classical mechanics (save in the gravitational regime). If that framework as a whole is to be understood inferentially, physics *as a whole* seems to be an inferential framework, and it is no longer clear what we are inferring *about*.

As for the dynamical conception, probability is at least a very different problem for QM than for classical SM. In the latter case, the problem was that "probabilities" were just an additional layer placed over an actual, deterministic, underlying dynamics, and that as such it was very hard to understand them as actually representing an objective property of the physical system. In unmodified QM, probabilities are features of the quantum state in certain decoherent regimes, and there is no additional "underlying dynamics" beyond the dynamics of the quantum state. In the case of dynamical collapse theories, or of hidden-variable theories with stochastic hidden-variable dynamics, probabilities are instead the result of genuinely stochastic laws. It is, of course, entirely possible to worry about the conceptual status of either explication of probability – Maybe we just can't make sense of probabilities in an Everett-type theory? Maybe stochastic laws don't make sense? – but it is at least clear what the analysis of probability would have to deliver, and clear that probability (whatever else it might be) is not an epiphenomenal gloss on underlying physics (there is, in other words, no way of separating the theory into probabilistic and non-probabilistic parts). Only in the special case of deterministic hidden-variable theories do we see the problems of classical-statistical-mechanical probability reproduced in a basically unchanged form.

Conversely, introduction of genuine stochasticity into classical statistical mechanics, as advocated (as a fundamental presumption) by Prigogine (1984) or (via quantum mechanics) by Albert (2000), would in effect reduce the problems of statistical-mechanical probability to the problems, if any, of understanding probability in dynamical-collapse theories. Such strategies have not been widely explored: for reasons not fully obvious to me, there seems to be more deference to unmodified classical mechanics in the light of the foundational problems of SM than to unmodified quantum mechanics in the light of the measurement problem.

4 Quantum Statistical Mechanics

So far I have drawn parallels between quantum and statistical mechanics, but the "statistical mechanics" I have considered is *classical* SM. Since the world is quantum rather than classical, however, presumably the project of understanding the applicability of SM to observed facts in the world should proceed via quantum rather than classical SM insofar as the two differ. Put another way, the project of studying even "classical" SM is, insofar as we want to understand the applicability of SM to the actual world, a project of studying classical SM as a limit or approximation to quantum SM rather than a project of studying classical SM in its own right. There is a different project, that of considering the logical coherence of classical SM in an exactly classical world; this is not the project I consider here (I will be concerned primarily with approaches to quantum theory that do not modify or supplement the baseline quantum formalism; I suspect much of my conclusion will carry over to dynamical-collapse or hidden-variable theories, but I leave that question to those more expert in those theories).

It is superficially tempting to suppose that we can map classical to quantum SM via a straight translation scheme: replace "phase space" by "(projective) Hilbert space," "phase-space point" by "Hilbert-space ray," and "probability distribution over phase space" by "probability distribution over projective Hilbert space," and then just apply the ideas of classical SM, mutatis mutandis. But I will argue that this is indefensible on a number of grounds.

To begin with a purely conceptual objection: classical micro-states are in no way probabilistic, but on either the inferential or the dynamical conception of quantum physics, there is a probabilistic aspect to the quantum state, so that a probability distribution over quantum states is in a sense a probability distribution over probability distributions. This is particularly stark if we take an inferentialist approach to QM, and regard the quantum state as in some sense epistemic. In this case, only an inferentialist approach to *statistical* mechanics seems to make sense, but then the probability distribution of SM is an epistemically interpreted probability distribution over epistemically interpreted probability distributions, and we might as well cut out the intermediate step. That is, if we take an inferentialist attitude to both quantum and statistical mechanics, the subject matter of the two disciplines is the same: both are concerned with our epistemic state. We will shortly see how this plays out technically.

Matters are somewhat more satisfactory *conceptually* if we take the dynamical conception of QM, and regard the quantum state as representing the physical world. We now have a choice of interpreting the probability distribution *over* quantum states as either an objective feature of those systems, or as epistemic. The former is at least inelegant: it requires two *separate* conceptions of objective probability – one to be understood via QM, the other via some unknown process. The latter is simplest to understand conceptually – we treat QM, like classical mechanics, as the underlying dynamical theory, and regard statistical-mechanical probabilities as quantifying our ignorance of the dynamical state in each case (for what it is worth, I have the impression that most authors commenting on the foundations of quantum SM think of it in this way).

However, even if the two notions of probability in play are conceptually distinct, they inevitably merge in any attempt to extract empirical content from the theory. Suppose we assign a probability measure $\Pr(\psi)$ over quantum states ψ. Then the expectation value of any observable $\hat{X}$ is given by

$$\langle \hat{X} \rangle = \int d\psi \; \Pr(\psi) \; \langle \psi | X | \psi \rangle. \tag{3}$$

As is well known, if we define the mixed state $\rho[\Pr]$ by

$$\rho[\Pr] = \int d\psi \; \Pr(\psi) \; |\psi\rangle\langle\psi|, \tag{4}$$

this expression can be rewritten as

$$\langle \hat{X} \rangle = \mathrm{Tr}\big(\rho[\Pr]\hat{X}\big). \tag{5}$$

So all empirical predictions about the system are determined directly by the mixed state, and only indirectly by the probability distribution. This matters because the relation between probability distributions and mixed states is many-to-one. Even if there are two notions of probability present (one emergent from the quantum dynamics, one entering through SM), they are mixed together in a way that defies empirical separation, once the initial method of preparation is lost.

Why does this matter? After all, even if the "correct" way of decomposing a mixed state is inaccessible to us, only crude instrumentalism would force us to deny that there is such a correct way. It matters because our evidence that quantum SM works comes exclusively through the empirical adequacy of expressions written down in terms of mixed states (whether those expressions are equilibrium equations of state, or non-equilibrium dynamical equations), and those expressions make contact with experiment in turn purely through the standard probability rule. So since the decompositional structure of a quantum state plays no role in those expressions, it is far from obvious that empirical data can tell us anything in general about the statistical micro-structure of a mixture of pure states. Indeed (since quantum SM itself, as a discipline, is formulated directly in terms of these mixed states), no amount of *empirical or theoretical* study of the theory looks well suited to tell us what these supposed "probability measures" are probability measures over.

Things become more dramatic still when we recall that so far we are working with a very impoverished notion of quantum state. As is well known, quantum systems can become entangled with their surroundings, and in doing so there is *no* Hilbert-space ray that correctly represents the system. Furthermore, the systems typically studied by SM are macroscopically large, so that there is absolutely no reason to expect that such systems, even if initially prepared in pure states, will not rapidly become entangled with their environment. That is, we have excellent dynamical grounds to expect that the probability of the system initially being in *any* pure state, even if it is initially taken to be 1, will rapidly approach zero.

There is a straightforward solution, of course: we can represent entangled systems by *mixed* states (by taking the partial trace over the environment of the state of the combined system). So if we want to hold on to the idea of placing a probability measure over quantum states, that measure had best be defined over the space of mixed states (in which pure states are present only as a special case).

If such a measure $\Pr(\sigma)$ is defined over mixed states σ, though, the many-to-one issue recurs in even sharper form. For all empirical predictions of the system are now given by the mixed state

$$\rho[\Pr] = \int d\sigma \; \Pr(\sigma) \; \sigma. \tag{6}$$

Any empirical result achieved by assigning to the system a *probability distribution* over mixed states can be reproduced by assigning it a *single* mixed state.

The introduction of probability measures over quantum states is thus epiphenomenal: any empirical prediction obtained via such a measure can be equally well obtained by assigning a single mixed state to the system – and on the dynamical conception of QM (and given entanglement), mixed states are perfectly legitimate choices of state for a physical system. If we want to interpret quantum statistical mechanics, we can do so simply by interpreting it in terms of mixed states understood primitively, without any additional need for a probability mixture over those states.

Couldn't we hold on to the idea that physical systems only have pure states by rejecting the idea that entangled systems have states at all? But there is no reason to think that even the *entire observable Universe* has a pure state. Indeed, there are rather good reasons to think that it does not, since the Universe as a whole has a causal horizon and in quantum field theory these horizons generally give rise to thermal radiation, the state of which is mixed.

So: on the dynamical conception of *quantum* mechanics, there seems simply no need to introduce additional probabilities via *statistical* mechanics, whether those probabilities are to be understood epistemically or in some more objective sense. The probabilities of quantum theory itself will do just fine.

It might be helpful at this point to compare QM with classical physics. In the latter, the "mixed states" are measures, or functions, over phase space. But these states (a) do not arise from classical micro-dynamics without explicit introduction of probability, and (b) are uniquely describable as probability measures over classical micro-states. So there is a fairly direct route, given the need for such states in classical SM, to infer that classical SM needs an additional probabilistic ingredient not present in the micro-physics. In QM, neither (a) nor (b) is true. If there is a need for additional probabilities in QM, then, an additional, entirely foundational argument is needed; to the best of my knowledge, no argument directly in terms of quantum SM (rather than an argument via classical analogy) has really been given.

We have already seen that the same ought to be true, on conceptual grounds, if we adopt the inferential conception of QM. We can now see technically how this goes through: on the inferential conception there is even less reason to deny that a mixed state is a legitimate state of a system (indeed, Gleason's theorem might be interpreted as telling us exactly why it is the most general such state). A probability distribution over mixed states is then just a probability distribution over probability distributions, and should collapse to a single mixed state, as indeed it does.

The conclusion seems to go through on either conception. While *classically* it might have seemed that classical SM is a probabilistic *generalization* of classical mechanics proper, quantum mechanically it is just a *restriction* of quantum mechanics to the regime in which statistical-mechanical methods (irreversibility, equilibrium, and the like) apply. *Which* regime that is depends on whether the dynamical or inferential conception is adopted. On the former, it is the regime in which the methods used to derive irreversible *dynamics* are applicable, and so is characterized inter alia by a large number of degrees of freedom and by an initial boundary condition that breaks the time symmetry (this holds for most unitary versions of quantum theory, at any rate; in stochastic modifications of quantum mechanics, the time asymmetry might arise from the time-asymmetric stochastic dynamics; see Albert 2000 and Wallace 2014 for further thoughts along these lines). On the latter, I have no idea, unless it is essentially characterized by external human intervention.

As a corollary, the questions of whether to adopt an inferential or a dynamical conception of physics cannot be answered independently in quantum and in statistical mechanics. The answer to the former determines the answer to the latter. And since classical mechanics is valid in our Universe only insofar as it is a valid approximation to quantum mechanics, these conclusions continue to hold true even in so-called classical statistical mechanics, which should be understood as quantum statistical mechanics in the classical limit. In particular, the probability distributions of classical statistical mechanics are quantum-mechanical states (pure or mixed) in a certain limiting regime (on this point, see Ballentine's observation (1990) that the classical limit of quantum mechanics is classical *statistical* mechanics; see also Uducec, Wiebe, and Emerson 2012, and my own observations in Wallace 2016a).

As a case study of how QM and quantum statistical mechanics have essentially the same subject matter, consider again the direction of time. Both in non-equilibrium statistical mechanics (in, e.g., the Boltzmann equation) and in decoherence theory, physicists are in the business of deriving time-asymmetric dynamical equations from a time-symmetric starting point (in both cases it is less than clear how to understand this from an inferential perspective). We might expect, then, that there is no sharp divide between the two sorts of derivation. Indeed, this turns out to be the case, as any perusal of the technical results in the respective fields illustrates. For instance, the decoherence master equation (a standard workhorse of environment-induced decoherence; see, e.g., Schlosshauer 2007) is derived by standard methods used in statistical physics to study dissipation; indeed, transformed to a phase-space representation the equation can be identified as simply the Fokker-Planck equation, a standard equation in kinetic theory (see, e.g., Liboff 2003: 301).

5 Conclusions and Considerations for Future Work

I have attempted to show that

(i) we can attempt to understand classical statistical mechanics on either the inferential or the dynamical conception, where in either case the underlying classical micro-state mechanics is understood dynamically;

(ii) we can likewise attempt to solve the measurement problem in quantum mechanics according to either conception, although the (apparent) absence of any underlying dynamical theory separate from quantum theory significantly alters the debate;

(iii) since quantum statistical mechanics can be understood as studying, directly, the quantum-mechanical states of individual systems (understood either inferentially or dynamically), the decision as to whether to understand quantum theory inferentially or dynamically forces the issue as regards the correct understanding of statistical mechanics.

The broader implications of these results for foundational and philosophical work lie beyond the scope of this chapter, but here I make some suggestions for future work:

- The current debate in the foundations of statistical mechanics, often characterized as "Gibbs vs. Boltzmann," might for some purposes be better characterized as "inferential vs. dynamical," as a respectable fraction of the criticisms made of the Gibbsian approach by Boltzmannians are really criticisms of the inferential conception rather than of the Gibbsian *machinery*.
- The quantum measurement problem can helpfully be understood as the problem of resolving a conceptual incoherence between probabilistic and non-probabilistic ways of understanding the quantum state, such that inferential (or ψ-epistemic) and dynamical (or ψ-ontic) approaches are different ways of resolving the incoherence: the former by treating the quantum state (somehow) as inherently representing an agent's probability function, the latter by treating probabilities as (somehow) emergent from a non-probabilistic underlying reality. Traditional ways of phrasing the measurement problem beg the question in favor of the dynamical conception and hinder communication. (This can still be the case even if the dynamical conception is *right*.)
- Although advocates of a ψ-epistemic view of the quantum state often advance their view by analogy to the classical probability distributions of statistical mechanics, this presupposes a controversial interpretation of classical statistical mechanics that in the quantum case collapses to a straightforward *restatement* of the ψ-epistemic view. The analogy ought therefore to be treated with rather more caution than is sometimes the case.
- While it may be of historical interest to understand how probabilities could have been understood in classical statistical mechanics considered in isolation, there is no point in seeking such understanding if our goal is to understand statistical mechanics in the actual world. In our world, the probabilities of statistical mechanics appear to be just special cases of the probabilities of *quantum* mechanics.

- The only exception (I can see) to the above concerns the de Broglie-Bohm theory. Here it is commonly claimed that probability is to be understood just as in classical statistical mechanics. Advocates of the theory thus have a clear need (and thus a clear motivation) to explore the interpretation and justification of probability in this context; they should not, however, be reassured by the supposed "fact" that ultimately they can piggy-back on the classical-statistical-mechanical explanation for such probabilities, as for all we know there may not be one.

- Although it is a common strategy (frequently adopted, less frequently defended) to study the foundations of *classical* statistical mechanics on the expectation that essentially the same issues arise in quantum statistical mechanics, this strategy is much more problematic than is often realized.

- In particular, the idea that quantum statistical mechanics involves putting probability distributions over quantum states (just as classical statistical mechanics involves putting probability distributions over classical states) appears very hard to justify.

I will conclude on a conciliatory note. Although it is probably apparent that I am much more sympathetic to the dynamical than to the inferential conception of quantum mechanics, and of statistical mechanics, *in general*, this need not mean that some *areas* of these fields are much better understood on something more like the inferential conception. In quantum theory, it is fairly clear that information theory ought to be understood that way: its subject matter is not the dynamical behavior of unattended systems, but the limitations imposed by physics on agents' activities. And in statistical mechanics, though the conceptual problems of statistical mechanics are often simply taken to be providing a micro-physical foundation for thermodynamics, "thermodynamics" is a misnomer: it is again concerned with agents' ability to control and transform systems, not primarily with how those systems behave if left to themselves.

References

Albert, D. Z. (2000). *Time and Chance*. Cambridge, MA: Harvard University Press.
Ballentine, L. E. (1970). "The Statistical Interpretation of Quantum Mechanics." *Reviews of Modern Physics*, **42**: 358–381.
Ballentine, L. E. (1990). *Quantum Mechanics*. Englewood Cliffs, NJ: Prentice Hall.
Brown, H. R., Myrvold, W., and Uffink, J. (2009). "Boltzmann's H-theorem, Its Discontents, and the Birth of Statistical Mechanics." *Studies in History and Philosophy of Modern Physics*, **40**: 174–191.
Bub, J. (1997). *Interpreting the Quantum World*. Cambridge: Cambridge University Press.
Callender, C. (2001). "Taking Thermodynamics Too Seriously." *Studies in the History and Philosophy of Modern Physics*, **32**: 539–553.

Caves, C., Fuchs, C., Manne, K., and Renes, J. (2004). "Gleason-Type Derivations of the Quantum Probability Rule for Generalized Measurements." *Foundations of Physics*, **34**: 193–209.

Fuchs, C. (2002). "Quantum Mechanics as Quantum Information (and Only a Little More)." Pp. 463–543 in A. Khrennikov (ed.), *Quantum Theory: Reconsideration of Foundations*. Växjö: Växjö University Press. Available online at http://arXiv.org/abs/quant-ph/0205039.

Gell-Mann, M. and Hartle, J. B. (1993). "Classical Equations for Quantum Systems." *Physical Review D*, **47**: 3345–3382.

Ghirardi, G., Rimini, A., and Weber, T. (1986). "Unified Dynamics for Micro and Macro Systems." *Physical Review D*, **34**: 470–491.

Gleason, A. (1957). "Measures on the Closed Subspaces of a Hilbert Space." *Journal of Mathematics and Mechanics*, **6**: 885–893.

Goldstein, S. (2001). "Boltzmann's Approach to Statistical Mechanics." Pp. 39–54 in J. Bricmont, D. Dürr, M. Galavotti, F. Petruccione, and N. Zanghi (eds.), *Chance in Physics: Foundations and Perspectives*. Berlin: Springer. Available online at http://arxiv.org/abs/cond-mat/0105242.

Greaves, H. (2007). "Probability in the Everett Interpretation." *Philosophy Compass*, **38**: 120–152.

Griffiths, R. B. (1984). "Consistent Histories and the Interpretation of Quantum Mechanics." *Journal of Statistical Physics*, **36**: 219–272.

Griffiths, R. B. (1996). "Consistent Histories and Quantum Reasoning." *Physical Review A*, **54**: 2759–2773.

Griffiths, R. B. (2002). *Consistent Quantum Theory*. Cambridge: Cambridge University Press.

Hajek, A. (1996). "'Mises Redux' – Redux – Fifteen Arguments against Finite Frequentism." *Erkenntnis*, **45**: 209–227.

Hajek, A. (2009). "Fifteen Arguments against Hypothetical Frequentism." *Erkenntnis*, **70**: 211–235.

Halliwell, J. J. (1998). "Decoherent Histories and Hydrodynamic Equations." *Physical Review D*, **35**: 105015.

Halliwell, J. J. (2010). "Macroscopic Superpositions, Decoherent Histories and the Emergence of Hydrodynamic Behaviour." Pp. 99–120 in S. Saunders, J. Barrett, A. Kent, and D. Wallace (eds.), *Many Worlds? Everett, Quantum Theory, and Reality*. Oxford: Oxford University Press.

Harrigan, N. and Spekkens, R. W. (2010). "Einstein, Incompleteness, and the Epistemic View of States." *Foundations of Physics*, **40**: 125–157.

Hartle, J. (2010). "Quasiclassical Realms." Pp. 73–98 in S. Saunders, J. Barrett, A. Kent, and D. Wallace (eds.), *Many Worlds? Everett, Quantum Theory, and Reality*. Oxford: Oxford University Press.

Hawthorne, J. (2010). "A Metaphysician Looks at the Everett Interpretation." Pp. 144–153 in S. Saunders, J. Barrett, A. Kent, and D. Wallace (eds.), *Many Worlds? Everett, Quantum Theory, and Reality*. Oxford: Oxford University Press.

Jaynes, E. (1957a). "Information Theory and Statistical Mechanics." *Physical Review*, **106**: 620–630.

Jaynes, E. (1957b). "Information Theory and Statistical Mechanics II." *Physical Review*, **108**: 171–190.

Jaynes, E. (1965). "Gibbs vs Boltzmann Entropies." *American Journal of Physics*, **5**: 391–398.

Joos, E., Zeh, H. D., Kiefer, C., Giulini, D., Kupsch, J., and Stamatescu, I. O. (2003). *Decoherence and the Appearance of a Classical World in Quantum Theory* (2nd Edition). Berlin: Springer.

Lavis, D. A. (2005). "Boltzmann and Gibbs: An Attempted Reconciliation." *Studies in the History and Philosophy of Modern Physics*, **36**: 245–273.

Lebowitz, J. (2007). "From Time-Symmetric Microscopic Dynamics to Time-Asymmetric Macroscopic Behavior: An Overview." Available online at http://arxiv.org/abs/0709.0724.

Liboff, R. L. (2003). *Kinetic Theory: Classical, Quantum, and Relativistic Descriptions* (3rd Edition). New York: Springer-Verlag.

Malament, D. and Zabell, S. L. (1980). "Why Gibbs Phase Space Averages Work: The Role of Ergodic Theory." *Philosophy of Science*, **47**: 339–349.

Maroney, O. (2012). "How Statistical Are Quantum States?" Available online at http://arxiv.org/abs/1207.6906.

Maudlin, T. (2010). "Can the World be Only Wavefunction?" Pp. 121–143 in S. Saunders, J. Barrett, A. Kent, and D. Wallace (eds.), *Many Worlds? Everett, Quantum Theory, and Reality*. Oxford: Oxford University Press.

Myrvold, W. (2016). "Probabilities in Statistical Mechanics." Pp. 573–600 in C. Hitchcock and A. Hájek (eds.), *Oxford Handbook of Probability and Philosophy*. Oxford: Oxford University Press.

Ney, A. and Albert, D. (eds.) (2013). *The Wave Function: Essays on the Metaphysics of Quantum Mechanics*. Oxford: Oxford University Press.

Omnés, R. (1988). "Logical Reformulation of Quantum Mechanics. I. Foundations." *Journal of Statistical Physics*, **53**: 893–932.

Omnés, R. (1992). "Consistent Interpretations of Quantum Mechanics." *Reviews of Modern Physics*, **64**: 339–382.

Omnés, R. (1994). *The Interpretation of Quantum Mechanics*. Princeton, NJ: Princeton University Press.

Pearle, P. (1989). "Combining Stochastic Dynamical State-Vector Reduction with Spontaneous Localization." *Physical Review A*, **39**: 2277–2289.

Peres, A. (1993). *Quantum Theory: Concepts and Methods*. Dordrecht: Kluwer Academic Publishers.

Prigogine, I. (1984). *Order Out of Chaos*. New York: Bantam Books.

Pusey, M. F., Barrett, J., and Rudolph, T. (2011). "On the Reality of the Quantum State." *Nature Physics*, **8**: 476–479. Available online at arXiv:1111.3328v2.

Russell, B. (1919). *Introduction to Mathematical Philosophy*. New York: Routledge.

Saunders, S., Barrett, J., Kent, A., and Wallace, D. (eds.) (2010). *Many Worlds? Everett, Quantum Theory, and Reality*. Oxford: Oxford University Press.

Schlosshauer, M. (2007). *Decoherence and the Quantum-to-Classical Transition*. Berlin: Springer.

Sklar, L. (1993). *Physics and Chance: Philosophical Issues in the Foundations of Statistical Mechanics*. Cambridge: Cambridge University Press.

Spekkens, R. W. (2007). "In Defense of the Epistemic View of Quantum States: A Toy Theory." *Physical Review A*, **75**: 032110.

Timpson, C. (2008). "Quantum Bayesianism: A Study." *Studies in the History and Philosophy of Modern Physics*, **39**: 579–609.

Tolman, R. C. (1938). *The Principles of Statistical Mechanics*. Oxford: Oxford University Press.

Uducec, C., Wiebe, N., and Emerson, J. (2012). "Equilibration of Measurement Statistics under Complex Dynamics." Available online at http://arxiv.org/abs/1208.3419.

Uffink, J. (2007). "Compendium of the Foundations of Classical Statistical Physics." Pp. 923–1074 in J. Butterfield and J. Earman (eds.), *Handbook for Philosophy of Physics*. Amsterdam: Elsevier.

Vranas, P. B. M. (1998). "Epsilon-Ergodicity and the Success of Equilibrium Statistical Mechanics." *Philosophy of Science*, **65**: 688–708.

Wallace, D. (2010a). "Decoherence and Ontology, or: How I learned to Stop Worrying and Love FAPP." Pp. 53–72 in S. Saunders, J. Barrett, A. Kent, and D. Wallace (eds.), *Many Worlds? Everett, Quantum Theory, and Reality*. Oxford: Oxford University Press.

Wallace, D. (2010b). "The Logic of the Past Hypothesis." Available online at http://philsci-archive.pitt.edu/8894/.

Wallace, D. (2012a). "Decoherence and Its Role in the Modern Measurement Problem." *Philosophical Transactions of the Royal Society A*, **370**: 4576–4593.

Wallace, D. (2012b). *The Emergent Multiverse: Quantum Theory According to the Everett Interpretation*. Oxford: Oxford University Press.

Wallace, D. (2014). "Probability in Physics: Statistical, Stochastic, Quantum." Pp. 194–220 in A. Wilson (ed.), *Chance and Temporal Asymmetry*. Oxford: Oxford University Press.

Wallace, D. (2015). "The Quantitative Content of Statistical Mechanics." *Studies in History and Philosophy of Modern Physics*, **52**: 285–293.

Wallace, D. (2016a). "Probability in Modern Statistical Mechanics: Classical and Quantum." Forthcoming.

Wallace, D. (2016b). "Reconciling Gibbsian and Boltzmannian Statistical Mechanics from a Dynamical Perspective." Forthcoming.

Zeh, H. D. (1993). "There Are no Quantum Jumps, nor Are There Particles!" *Physics Letters A*, **172**: 189–192.

Zurek, W. H. (1991). "Decoherence and the Transition from Quantum to Classical." *Physics Today*, **43**: 36–44. Revised version available online at http://arxiv.org/abs/quant-ph/0306072.

Zurek, W. H. (1998). "Decoherence, Einselection, and the Quantum Origins of the Classical: The Rough Guide." *Philosophical Transactions of the Royal Society of London A*, **356**: 1793–1820. Available online at http://arxiv.org./abs/quant-ph/98050.

Zurek, W. H. (2003). "Decoherence, Einselection, and the Quantum Origins of the Classical." *Reviews of Modern Physics*, **75**: 715–775.

Zwanzig, R. (1961). "Memory Effects in Irreversible Thermodynamics." *Physical Review*, **124**: 983–992.

10

Classical Models for Quantum Information

FEDERICO HOLIK AND GUSTAVO MARTÍN BOSYK

1 Introduction

The past decades have seen an impressive development of quantum information theory (see, e.g., Bennett 1995; Nielsen and Chuang 2010). While the construction of quantum computers remains challenging (Di Vicenzo 1995; Steane 1998; Deutsch and Ekert 2000), other important advances have been achieved (Carolan et al. 2015). Furthermore, new technologies have been applied in communicational security, and many quantum information protocols have been reproduced in the lab.

But besides the technological innovations and promises, the development of quantum information techniques has given rise to a (maybe unexpected) rebirth of deep foundational questions (see, e.g., Clifton et al. 2003; Amaral et al. 2014; Popescu 2014). The foundations of quantum mechanics are now seen in a new light. And a new language is used to interrogate our universe, which supplies us knowledge about what nature allows us to do and what not with regard to the transmission and processing of information. What are the physical limits for the transmission of information? Which is the role played by locality and no-signaling conditions in this scenario? What can be computed and what cannot by appealing to realistic physical devices? These are the kind of questions researchers have been wondering, and it turned out that the task of answering them is not trivial at all (see, e.g., Steane 1998; Deutsch and Ekert 2000; Brunner et al. 2014; Popescu 2014). In this setting, the quest for the fundamental principles that specify quantum correlations among other possibilities has played a central role (Popescu 2014). This quest led, for example, to the formulation of the Information Causality Principle (Barnum et al. 2010) and the Exclusivity Principle (Gühne et al. 2010; Amaral et al. 2014; Cabello 2014), among others (Clifton et al. 2003; Oas and de Barros 2015).

While there has been an impressive development on the technical side, many debates remain about the meaning of the informational approach for the foundations of quantum mechanics. Specifically: what is quantum information? Can we give a coherent meaning to such a notion? These questions, and even the physical status of the very notion of "quantum information," have been largely debated (see Lombardi et al. 2016a, and references therein). There are other alternatives at hand: for example, in the past years, quantum Bayesianism has grown substantially as a subjective attitude toward these problems (Fuchs and Shack 2013). But, after many years, there seems to be no agreement regarding the interpretation of the informational approach and the meaning of the notion of quantum information.

In this chapter, we will discuss the ontological status of the notion of quantum information. In the first part, we review the mathematical framework of a generalized information theory. In this framework, probability theory will play a central role and, thus, we will devote Section 2 to it. We argue that the existence of probabilistic models that go beyond the classical and quantum realms, and the possibility of performing informational protocols in those models, allow us to claim that a generalized information theory can be conceived. Notice that this perspective is complementary to that of other research programs (see, e.g., Holik et al. 2015; Holik et al. 2016, and references therein).

In the second part, we will consider the question about the ontology of the possible models satisfying the different instances of that generalized formalism. This will be the content of Section 3. Specifically, we want to study the role quantum mechanics plays in quantum information protocols; the question is whether there are other possible ontologies that can perform the same tasks. We will relate this problem to the existence of many examples of physical systems that are actually constructed using an essentially classical ontology but, on the other hand, are modeled by structures resembling the laws of quantum mechanics (e.g. Aerts 1998; Kwiat et al. 2000; Spreeuw 2001; Couder et al. 2005; Couder and Fort 2006; Qian et al. 2015). Their significance for us relies in the fact that they can be used to perform quantum information protocols (though, in an inefficient way) (Man'ko et al. 2001; Bhattacharya et al. 2002; Francisco et al. 2006a, 2006b; Francisco and Ledesma 2008; Goldin et al. 2010). On this basis, we will explore the ontological implications of the existence of these simulations. Finally, in Section 4 we will draw our conclusions.

2 A General Formalism

In this section, we describe the mathematical formalism of information theory. We will stress the relationship between the laws governing the underlying probabilities of each theory and the main features of the concomitant information theory. We end up by arguing that the existence of generalized probabilistic models allows us to talk of a generalized information theory (see Holik et al. 2015 for further discussion).

2.1 Classical Probabilities

In the seminal paper on classical information theory, Claude Shannon (1948: 379) stated that:

A basis for such a theory [classical information theory] is contained in the important papers due to Nyquist and Hartley on this subject. In the present paper we will extend the theory to include a number of new factors, in particular the effect of noise in the channel, and the savings possible due to the statistical structure of the original message and due to the nature of the final destination of the information.

The reference to the "statistical structure of the original message" anticipates for the reader that probabilitics will play a key role in Shannon's formalism. But what is the nature of the probabilities involved in Shannon's formalism? In order to answer this question, we will begin by discussing possible axiomatizations for probability theory. In this chapter, we will show that classical probabilities are involved in Shannon's theory. But if non-classical measures are used, a different information theory arises.

Perhaps, the most widely accepted list of axioms for classical probabilities is that of Andrey Kolmogorov (1933). Let Σ be a σ-algebra of events and μ a probability measure that assigns a real number to every subset in Σ, such that:

$$\mu(A) \geq 0 \quad \text{for all } A \text{ in } \Sigma \tag{1}$$

$$\mu(\Sigma) = 1 \tag{2}$$

$$\mu(\bigcup_i A_i) = \sum_i \mu(A_i) \text{ if all } A_i \text{ are mutually exclusive.} \tag{3}$$

As a consequence, one has

$$0 \leq \mu(A) \leq 1 \tag{4}$$

$$\mu(\varnothing) = 0 \tag{5}$$

$$\mu(A^c) = 1 - A^c, \tag{6}$$

Where "$(\ldots)^c$" denotes the set-theoretical complement. The probabilities used in Shannon's formalism can be described using Kolmogorov's axioms.

There are of course other approaches to probability theory. Some of the most important alternatives are those of Bruno de Finetti (1937) and Richard Threlkeld Cox (1946, 1961). Cox's approach has been also extended to the quantum domain (see, e.g., Holik et al. 2014, and references therein). The formalism can also be extended in order to include negative probabilities (de Barros and Oas 2014). In the following sections, we will discuss more general probabilistic measures.

2.2 Probabilities in Quantum Mechanics

Regarding the probabilities involved in quantum theory, Richard Feynman (1951: 533) stated:

I should say, that in spite of the implication of the title of this talk, the concept of probability is not altered in quantum mechanics. When I say "the probability of a certain outcome of an experiment is p," I mean a conventional thing, that is, if the experiment is repeated many times one expects that the fraction of those which give the outcome in question is roughly p. I will not be at all concerned with analyzing or defining this concept in more detail, for no departure of the concept used in classical statistics is required. What is changed, and changed radically, is the method of calculating probabilities.

What does Feynman mean when he says that *the method of calculating probabilities* has changed radically? A clue to answer this question can be found in the paper of Christopher Fuchs and Rüdiger Schack (2013), where the authors claim:

The concept of probability is not altered in quantum mechanics (it is personalistic Bayesian probability). What is radical is the recipe it gives for calculating new probabilities from old.

Clearly, there are rules for calculating classical probabilities. As an example, suppose that we know that the probability of occurrence of an event A is $P(A)$, and that of event B is $P(B)$. With this information, how to compute the probability $P(A \vee B)$? If one wants to remain consistent with Kolmogorov's axioms, we must have $P(A \vee B) = P(A) + P(B) - P(A \wedge B)$. A similar analysis due to de Finetti justifies the use of the usual rules of classical probability calculus. How do these matters change in quantum mechanics?

In quantum mechanics, the state of the system is represented by a density operator ρ, which is positive definite and has trace one, i.e., $\rho \geq 0$ and $\mathrm{Tr}\,\rho = 1$. On the other hand, a quantum observable is represented by an Hermitian operator, i.e., $A = A^{\dagger}$, where "$\dagger$" denotes complex conjugation, whereas its eigenvalues a_i, are the possible outcomes of the measurement of A. Thus, the probability (in Feynman's sense) that an outcome a_i of an observable A is given by Born's rule:

$$P_{\rho}(a_i) = \mathrm{Tr}(\rho \mathrm{P}^i_A), \tag{7}$$

where P_A^i is the projector associate to the eigenvalue a_i. This notion can be extended to any quantum effect E: $P_\rho(E) = \mathrm{Tr}(\rho E)$. According to Gleason's theorem (1957), any probability assignment s of the form (7) can be represented by a density operator ρ that defines a measure s over the set of projectors $\mathcal{P}(\mathcal{H})$ as follows:

$$s(\mathrm{P}) \geq 0 \text{ for all } \mathrm{P} \text{ in } \mathcal{H} \tag{8}$$

$$s(1) = 1 \tag{9}$$

$$s\left(\sum_i \mathrm{P}_i\right) = \sum_i s(\mathrm{P}_i) \text{ if all } \mathrm{P}_i \text{ are mutually orthogonal.} \tag{10}$$

In other words, Gleason's theorem states that for each density operator ρ, there exists a unique measure s satisfying axioms (8)–(10) and vice versa; so, there is a one-to-one correspondence between them. This means that the right way to axiomatize probabilities in quantum mechanics is given by the axioms listed earlier; we will call them "von Neumann's axioms," in order to distinguish them from Kolmogorov's axioms.

Which are – despite their similarities – the main differences between Kolmogorov's and von Neumann's axioms? The first one is that Σ is a collection of subsets of a set of outcomes, whereas $\mathcal{P}(\mathcal{H})$ is a collection of projection operators acting on a Hilbert space. The second difference relies on the involved event algebras: Boolean in the classical case and non-Boolean in the quantum case. Indeed, $\mathcal{P}(\mathcal{H})$ fails to be distributive. Another important difference is that the negation in Σ is represented by the set theory complement, whereas negation in $\mathcal{P}(\mathcal{H})$ is represented by the orthogonal complement of projections (which is coincident with that of subspaces, because they can be put in one-to-one correspondence). Another important difference is given by the fact that, if the quantum logical connective "$\vee$" is used, then it may happen that $P(A \vee B) \geq P(A) + P(B)$ depending on the state and the propositions involved.

The non-Booleanity of $\mathcal{P}(\mathcal{H})$ has to do with the fact that complementarity plays a key role in quantum mechanics. It expresses the need of using complementary experimental setups which, while being mutually incompatible, are notwithstanding indispensable to provide a complete description of phenomena. This is at the basis of quantum contextuality (de Barros and Oas 2015). This means that the quantum formalism dictates that each empirical context will be represented by a Boolean subalgebra of $\mathcal{P}(\mathcal{H})$. But the complete description of the quantum system requires all Boolean subalgebras (empirical contexts), resulting in an overall non-Boolean algebra of events.

A quantum state defines a classical probability measure on each Boolean subalgebra of $\mathcal{P}(\mathcal{H})$. Thus, a quantum state could be considered as a collection of Kolmogorovian probability distributions (indexed by the empirical contexts) (Holik et al. 2015; Holik et al. 2016). But this collection is not arbitrary: *the quantum state pastes all these probability distributions together in a coherent way.*

Let us see how this works from a formal perspective. It is important to recall that an arbitrary orthomodular lattice $\mathcal{L}$ can be written as a disjunction:

$$\mathcal{L} = \bigvee_{\mathcal{B} \in B} \mathcal{B} \tag{11}$$

Where B is the set of all possible Boolean subalgebras $\mathcal{B}$ of $\mathcal{L}$. A state s on $\mathcal{L}$ defines a classical probability measure on each $\mathcal{B}$. In other words, $s_{\mathcal{B}}(\ldots) := s|\mathcal{B}(\ldots)$ is a Kolmogorovian measure over $\mathcal{B}$. Let us see some examples in order to show how this works. Notice that when $\mathcal{H}$ is finite dimensional, its maximal Boolean subalgebras will be finite.

Example 1: Let $\mathcal{P}(\mathbb{C}^2)$ be the lattice of projection operators of a two-dimensional quantum system or *quantum bit* (*qubit*). Then, a Boolean subalgebra has the form $\{0, \mathrm{P}_1, \mathrm{P}_2, 1_{\mathbb{C}^2}\}$ with $\mathrm{P}_i = |\varphi_i\rangle\langle\varphi_i|$ for some orthonormal basis $\{|\varphi_1\rangle, |\varphi_2\rangle\}$ (see Figure 10.1). Notice that each Boolean subalgebra is isomorphic to the Boolean algebra of two elements $\mathcal{B}_2 = \{1, 2\}$.

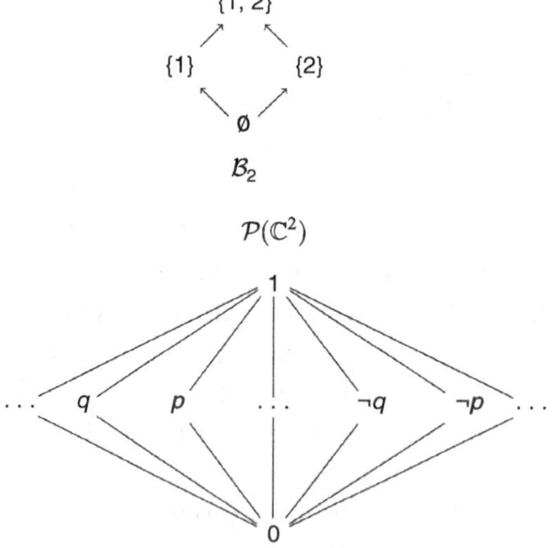

Figure 10.1. Boolean algebra of two elements and its quantum counterpart $\mathcal{P}(\mathbb{C}^2)$.

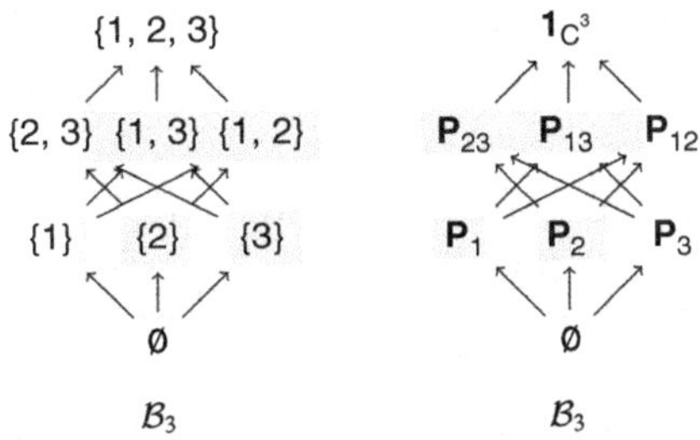

Figure 10.2. Boolean algebra of three elements and the Boolean subalgebra of $\mathcal{P}(\mathbb{C}^3)$ generated by three orthogonal one-dimensional projection operators.

Example 2: Let $\mathcal{P}(\mathbb{C}^3)$ be the lattice of projection operators of a three-dimensional quantum system or *quantum trit* (*qutrit*). Then, a Boolean subalgebra has the form $\{0, P_1, P_2, P_3, P_{12}, P_{13}, P_{23}, 1_{\mathbb{C}^3}\}$ with $P_i = |\varphi_i\rangle\langle\varphi_i|$ and $P_{ij} = |\varphi_i\rangle\langle\varphi_i| + |\varphi_j\rangle\langle\varphi_j| (i \neq j)$ for some orthonormal basis $\{|\varphi_1\rangle, |\varphi_2\rangle, |\varphi_3\rangle\}$ (see Figure 10.2). Notice that each Boolean subalgebra is isomorphic to the Boolean algebra of three elements, formed by all the possible subsets of $\mathcal{B}_3 = \{1, 2, 3\}$.

Summarizing, we can say that:

- A quantum state ρ defines a Kolmogorovian probability distribution on each maximal subalgebra of an orthomodular lattice.
- These Kolmogorovian measures are pasted together in a coherent way via Born's rule (or equivalently, due to von Neumann's axioms).

2.3 More General Contextual Models

Here, we will briefly present the general contextual models, which are defined by the following axiomatic. Let $\mathcal{L}$ be the lattice of all events, and s a measure over $\mathcal{L}$ such that:

$$s(E) \geq 0 \text{ for all } E \text{ in } \mathcal{L} \tag{12}$$

$$s(1) = 1 \tag{13}$$

$$s\left(\sum_i E_i\right) = \sum_i s(E_i) \text{ if all } E_i \text{ are mutually orthogonal.} \tag{14}$$

These measures do not exist for all possible orthomodular lattices (Beltrametti and Cassinelli 1981). Notice that Eqs. (1)–(3) and (8)–(10) can be viewed as particular cases of Eqs. (12)–(14).

It is possible to develop an extension of Cox's approach for quantum theory and generalized probabilistic models, in which the structure of axioms (12)–(14) acquire a clear operational meaning:

- According to Cox's approach to classical probability (1946, 1961), if a rational agent deals with a Boolean algebra of assertions representing physical events, a plausibility calculus can be derived in such a way that the plausibility function yields a theory that is formally equivalent to that of Kolmogorov.
- It can be proved that a similar result holds if the rational agent deals with an atomic orthomodular lattice representing a contextual theory (Holik et al. 2014). For the quantum case, non-Kolmogorovian measures arise as the only ones compatible with the non-commutative (non-Boolean) character of quantum complementarity.

2.4 Generalized Information Theory

We have seen that there exist theories (such as quantum mechanics) whose states define non-Boolean distributive measures. But we have also remarked that probabilities played a key role in Shannon's formalism. Which are the implications of the non-classicality of the underlying probabilistic axioms for information processing? In other words: how would our capabilities of processing information change if the state measures become non-classical?

Let us see first what happens with quantum mechanics. Shannon entropy plays a key role in coding a message: it can be considered as the mean optimal code length. In Cox's approach, Shannon's information measure relies on the axiomatic structure of Kolmogorovian probability theory. It can be shown that the von Neumann entropy arises as a natural measure of information derived from the non-Boolean character of the underlying lattice $\mathcal{P}(\mathcal{H})$ (Holik et al. 2016). On the basis of this fact, it is not surprising to find that it is the optimal bound for coding in a quantum channel (Schumacher 1995). From this formal perspective, Schumacher's coding theorem would be the natural non-Boolean version of Shannon's theorem.

In turn, classical information theory and quantum information theory can be considered as particular cases of a more general non-commutative or contextual information theory. Let us see how this works in the general case. It is easy to show that states satisfying axioms (12)–(14) form a convex set. The fact that the set of all possible states (we call it $\mathcal{C}$ from now on) is convex has a very natural interpretation: if s_1 and s_2 are two states, then, any convex combination of them, $\omega s_1 + (1 - \omega)s_2$ with $\omega \geq 0$, is a possible statistical mixture. Thus, a suitable setting for describing generalized probabilistic models is the one based on convex sets. In this generalized setting, it is possible to work with generalized events as follows (Holik and Plastino 2014). A given state s of the system should assign to any – generalized – event E a real number between zero and one, i.e., $s(E) \in [0, 1]$. This is a natural requirement if one aims at developing an interpretation similar to that of Feynman (see Section 2.2). Notice that any event E defines a functional over the set: $E(s) := s(E)$ for all $s \in C$. Assume, without loss of generality, that C can be embedded in a vector space V_C. Then, it should be clear that all possible events can be put into a one-to-one correspondence with a positive cone in the dual space V_C^*. As in the quantum and classical cases, it is reasonable to assume that a unit functional u exists such that $u(s) = 1$ for all $s \in C$ (this role is played by the trace functional and the identity function for the quantum and classical cases, respectively). A (discrete) observable A with outcomes $\{e_i\}$ can thus be defined as a collection of events $\{E_i\}$, such that $\sum_i E_i(\ldots) = u(\ldots)$. This is a clear generalization of quantum event, a notion that includes projective measurements as particular cases (Holik and Plastino 2014).

From the perspective described previously, a particular model or theory can be considered as a convex set C endowed with a dual set of observables and a normalization functional $u(\ldots)$. Can we speak of information processing in this general setting? The answer is affirmative, and there has been considerable work in this direction (see Holik and Plastino 2014; Holik et al. 2015 for a complete list of references). In particular, information measures can be studied in this setting (see Table 10.1). The natural information measure for a classical probabilistic model is given by Shannon entropy. In quantum mechanics, this measure is no longer suitable, and von Neumann entropy must be introduced. In the generalized setting, a *measurement entropy* arises (see Short and Wehner 2010 for details). These information measures can be endowed with a natural operational interpretation according to the extension of the Cox's approach to the generalized case (presented in Holik et al. 2016).

Table 10.1. *Information measures for classical, quantum, and general cases.*

	Classical	Quantum	General
Lattice	$\mathcal{P}(\Gamma)$	$\mathcal{P}(\mathcal{H})$	$\mathcal{L}$
Entropy	$-\sum_i p_i \ln p_i$	$-\mathrm{Tr}(\rho \ln \rho)$	$\inf_{F \in \mathcal{C}} H_F(u)$

3 Ontological Commitments

In the previous section, we have seen that a formal framework exists for a generalized information theory: an abstract formulation of information theory and informational notions written in an operational way. But what about ontology? Which is the physical or conceptual correlate of that formal framework? In particular, in which sense do we speak about information when we code something using qubits? For example, one may wonder whether it is possible to interpret a general string of qubits as a message. If their states form an orthonormal basis, we recover a classical notion of information (whatever it is). But if non-orthogonal states are used for coding, the question is subtler (see Lombardi et al. 2016b).

Rolf Landauer (1996) has pointed out that, due to the fact that abstract information must be always instantiated in concrete physical systems, then there should be a connection between physics and information theory. On this basis, there were many attempts to claim that information is a kind of physical quantity (such as electric current, electric capacity, entropy, or energy). In this way, it could be naturally transmitted, stored, or processed. But if we want to avoid the commitment with this kind of physical interpretation of information, we can consider that the content of a physical string of bit or of qubits is just its operational/computational task in the particular context that it is used.

But all these proposals are not sufficient to find a cogent ontological interpretation. Thus, we can ask: is quantum mechanics really necessary to implement the formalism of quantum information processing? If yes, in which sense? Which ontologies are compatible with the formal framework? In this section, we will first review the role of complementarity in quantum information processing. Then, we will discuss the possible ontologies compatible with the mathematical formalism. Finally, we will study the role of classical models of quantum information.

3.1 Complementarity and Information

We have seen that contextuality is an essential formal aspect in von Neumann's axioms for probabilities in quantum mechanics. As Kolmogorovian probabilities constitute a central aspect in classical information theory, it is to be expected that the contextual character of quantum probabilities be also important in quantum information protocols. And this is indeed the case. For example, it can be shown that the non-Boolean character of the lattice $\mathcal{P}(\mathcal{H})$ plays a key role in some important quantum computation protocols (see Bub 2007).

As Joseph Renes (2013) remarks, one major question in quantum information theory is to determine what happens to information processing when the carriers of information are described by the laws of quantum mechanics. Renes explores the role of complementarity in quantum information theory. He attempts to provide a deeper understanding of the theory by studying its role in many examples of quantum information protocols. In this context, he claims that (Renes 2013: 30):

This reflects our main theme that what really counts in quantum information is classical information about complementary observables.

One of the main conceptual tools that can be used to discuss the key concepts of Renes' article is the so-called information game. We will reproduce it here, because it illustrates the role of complementarity in quantum information without appealing to complex mathematics. The information game consists of two players, Alice and Bob. In the classical version of the game, Bob places a coin into a box and sends it to Alice. Next, Alice asks Bob if she will see heads or tails when she opens the box and looks at the state of the coin. Bob's goal is to predict Alice's outcome. This can be repeated many times. In order to win the game, Bob could write down on a piece of paper (or use any other dispositive serving as a memory) if he put the coin head or tail side up. If the memory device works correctly, he will be able to predict Alice's outcome with success. The success of Bob's predictions depends heavily on the accuracy with which the memory device stores the information.

It is clear that the piece of paper carries information about the state of the coin. But how much information? This is – essentially – the problem of Shannon. Renes explains it as follows. The state of the coin chosen by Bob can be represented by a binary random variable X, taking two values with probability p for heads and $1-p$ for tails (with whatever probability he decides). The state of the memory device can be represented by a random variable M, and the winning strategy requires $M = X$ for any choice of X. The amount of information stored in the memory can be naturally measured using Shannon entropy, which, for a probability distribution p, reads: $H(p) = -p\ln p - (1 - p)\ln(1 - p)$.

In the quantum version of the game, Bob prepares a qubit in a certain quantum state ρ and sends it to Alice. Alice asks Bob which will be the outcome when she makes a measurement on the qubit. She is allowed to use only two complementary observables, namely, amplitude Z and phase X. This can be illustrated by using the Mach-Zehnder interferometer (see Figure 10.3).

If an experiment is designed to measure which way of the interferometer (represented by the two states $|z_0\rangle$ and $|z_1\rangle$) was taken by the particle, it will be called an amplitude measurement Z. Its corresponding operator is written as $Z = |z_0\rangle\langle z_0| - |z_1\rangle\langle z_1|$. Contrarily, if interference is measured by placing a second beam splitter and two detectors at the end of the setup, it will be called a phase measurement. A phase measurement will be a projective measurement in the basis $|x_\pm\rangle = 1/\sqrt{2}(|z_0\rangle \pm |z_1\rangle)$. The corresponding observable is represented

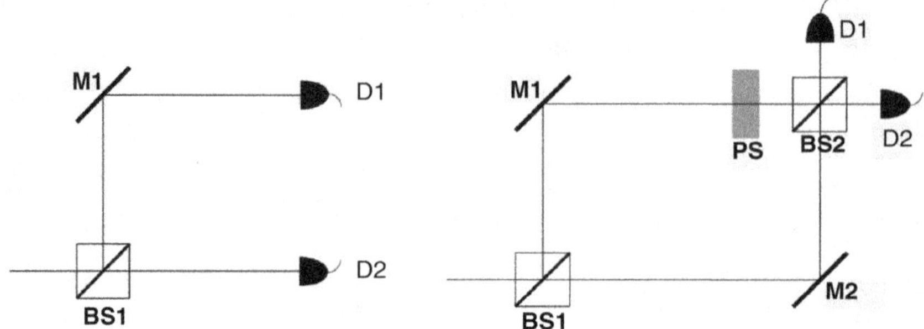

Figure 10.3. Left: experimental setup for performing a "which way" measurement. Right: experimental setup for performing an interference measurement.

by the operator $X = |x_+\rangle\langle x_+| - |x_-\rangle\langle x_-|$. Thus, the complementary observables X and Z satisfy the entropic uncertainty relation (Maassen and Uffink 1988):

$$H(X)_\rho + H(Z)_\rho \geq \ln 2 \tag{15}$$

where $H(X)_\rho$ and $H(Z)_\rho$ are the Shannon entropies of the corresponding statistics of the observables X and Z, respectively, for the given state ρ.

Now suppose that Alice is free to make either an amplitude or a phase measurement, but she does not tell Bob her choice. Bob is free to prepare any state he wants, but due to the complementarity captured by the entropic uncertainty relation (15), neither state has definite values for both observables. Then, any attempt to win the game with certainty is doomed to fail. On the contrary, if after her choices of measurement in a single run of the experiment Alice asks for a prediction of this particular measurement, there exists a remarkable winning strategy provided that Bob is allowed to use a quantum memory, that is, a second qubit entangled with Alice's qubit. Indeed, the trick is to store quantum information about the qubit that Bob sends to Alice! By using Bell states such as $|\Phi^{AB}\rangle = 1/\sqrt{2}(|z_0^A\rangle|z_0^A\rangle + |z_1^A\rangle|z_1^B\rangle) = 1/\sqrt{2}(|x_+^A\rangle|x_+^A\rangle + |x_-^A\rangle|x_-^B\rangle)$ and sending the qubit A to Alice, Bob can always win the modified version of the game by means of his quantum memory (qubit B). When Alice asks for the result of her particular choice of measurement, Bob can simply make the same measurement in his memory, and the results will match.

Thus, in a certain sense, the quantum memory instantiated by system B contains one bit of classical information about both the amplitude and phase of system A. Renes concludes that this combination of classical information about complementary contexts lies at the heart of the notion of quantum information. He also reminds us the remarkable fact that entangled states display correlations even if there is nothing there to correlate. While at first glance it would seem that these astonishing results contradict the uncertainty principle, this is not the case. Since Bob makes use of system B, then the entropies should be computed conditioned on this fact. This situation is correctly described by a recently introduced entropic uncertainty relation in the presence of quantum memory (Renes 2013),

$$S(X|B) + S(Z|B) \geq \ln 2 + S(A|B) \tag{16}$$

where $S(X|B)$, $S(Z|B)$ and $S(A|B)$ denote the conditional von Neumann entropies of states $(X \otimes I)\rho^{AB}$, $(Z \otimes I)\rho^{AB}$ and ρ^{AB}, respectively, and $S(A|B)$ is a signature of the degree of entanglement between systems A and B. The right-hand side of (16) vanishes for Bell states, trivializing the inequality. In this case, Bob could figure out the result of both experiments. On the other hand, this inequality reduces to usual entropic uncertainty relation (15) in the case that Bob has no quantum memory and prepares pure states.

As this discussion shows, complementarity plays a key role in quantum information protocols. Any physical device (classical or not) aiming at reproducing these protocols should also reproduce complementarity. In the following sections, we will consider ontological matters and discuss classical models of quantum systems, that is, classical systems which manage to imitate quantum contextual behavior. Then, we will extract the conclusions for the interpretation of a generalized quantum information theory.

3.2 Many Worlds and Algorithmic Complexity

Perhaps, the most radical interpretation of quantum information is that based on the idea of many worlds (Everett 1957). Let us explain it as follows. Consider a quantum memory with a three-qubit capacity. Then, there are eight elements in the computational basis: $\{|000\rangle, |100\rangle, |010\rangle, |110\rangle, |001\rangle, |101\rangle, |011\rangle, |111\rangle\}$. If we have had three classical bits, then, we would have been able to store only one number out of the eight possibilities, namely, $\{000, 100, 010, 110, 001, 101, 011, 111\}$. But in the quantum case, we can produce the following initial state:

$$\frac{1}{\sqrt{8}} \left(|000\rangle + |100\rangle + |010\rangle + |110\rangle + |001\rangle + |101\rangle + |011\rangle + |111\rangle \right) \tag{17}$$

Thus, this state seems to code all possible sequences at the same time in the same physical device. Furthermore, a unitary evolution representing a quantum logical gate will act simultaneously on all the elements of the superposition due to its linear character. In other words, it will perform a computation on each of the eight elements of the computational basis. In a sense, a quantum computation could be considered as a family of classical computations, each of them running in parallel, at the same time on the same system. Somehow, the quantum computer manages to work on all the inputs at the same time, as if it operated in parallel universes or realities (each one with a different input). This intuition was considered appealing by many physicists and computer scientists, and was developed into a full interpretation. For example, David Deutsch and Artur Ekert (2000: 96) put this as follows:

Thus quantum theory describes an enormously larger reality than the universe we observe around us. It turns out that this reality has the approximate structure of multiple variants of that universe, co-existing and affecting each other only through interference phenomena – but for the purposes of this article, all we need of these "parallel universes" ontology is the fact that what we see as a single particle is actually only one tiny aspect of a tremendously complex entity, the rest of which we cannot detect directly. Quantum computation is about making the invisible aspects of the particle – its counterparts in other universes – work for us.

And they continue:

This is our "principle of local operations": a single local unitary operation on a subsystem on a large entangled system processes the embodied information by an amount which would generally require an exponential effort to represent in classical computational terms. In the sense noted above, n qubits have exponentially larger capacity to represent information than n classical bits. However the potentially vast information embodied in a quantum state has a further remarkable feature – most of it is inaccessible to being read by any possible means!

In a similar vein, Richard Jozsa (2000: 109) states:

The full (largely inaccessible) information content of a given quantum state is called quantum information. Natural quantum physical evolution may be thought of as the processing of quantum information. Thus the viewpoint of computational complexity reveals a new bizarre distinction between classical and quantum physics: to perform natural quantum physical evolution, Nature must process vast amounts of information at a rate that cannot be matched in real time by any classical means, yet at the same time, most of this processed information is kept hidden from us! However it is important to point out that the inherent inaccessibility of quantum information does not cancel out the possibility of exploiting this massive information processing for useful computational purposes.

The first thing to notice is that, with these observations, the authors are providing an ontological interpretation for quantum information processing based on many worlds. But leaving aside this particular interpretation, it is very important to stop here and to think about which are the implications of the existence of these quantum algorithms.

The Extended Church-Turing Thesis can be stated as follows: *every finitely realizable physical system can be simulated arbitrarily closely by a universal model computing machine operating by finite means* (Aaronson and Arkhipov

2011). There are good reasons nowadays to believe that the Extended Church-Turing Thesis is false. Indeed, the developments on quantum computing theory during the past decades seem to indicate that computational tasks exist that cannot be solved efficiently by appealing to classical computers, but can be solved with polynomial resources using quantum computers (Aaronson and Arkhipov 2011). But the full-size realization of quantum computers is not a trivial task. In this context, some alternatives were proposed as intermediate quantum computational models that can be used to solve problems that are believed to be classically hard. In the following, we will assume (as nowadays physicists do) the supremacy of quantum computers over the classical ones.

3.3 *Classical Models of Quantum Mechanics*

Sandu Popescu (1994) showed that it is possible to implement the teleportation protocol with reasonable fidelity by using states that do not violate Bell's inequalities. This implies that the teleportation protocol can be implemented by means of states that can be modeled by local hidden variables. This result indicates that it is possible to conceive systems that are essentially classical, but that mimic quantum features, specifically, quantum information protocols.

Classical models imitating quantum behavior have been extensively studied in the literature. A simple example that allows us to illustrate their main features was introduced by Diederik Aerts and collaborators (see, e.g., Aerts 1998). There, a ball placed in the surface of a sphere in direction u, falls orthogonally into an elastic band at point c, as indicated in Figure 10.4 (a) and (b).

The elastic band breaks with uniform probability (Figure 10.4 (c)). The two pieces of the band contract each other and reach the surface of the sphere. The ball ends up at the point of the sphere corresponding to where the elastic breaks (Figure 10.4 (d)). If the elastic band is placed in direction v, the probabilities for the ball to reach v and $-v$, respectively, can be computed and are given by

$$p(v; u) = \cos^2(\theta/2) \tag{18}$$

$$p(-v; u) = \sin^2(\theta/2) \tag{19}$$

where θ is the angle between u and v. The elastic band model can be refined in order to describe improper mixtures, entanglement, and the transition to the classical limit (see Aerts 1998 for details). In this model, mixed states are described by locating the ball in any place of the sphere (it is easy to see that this makes the ball

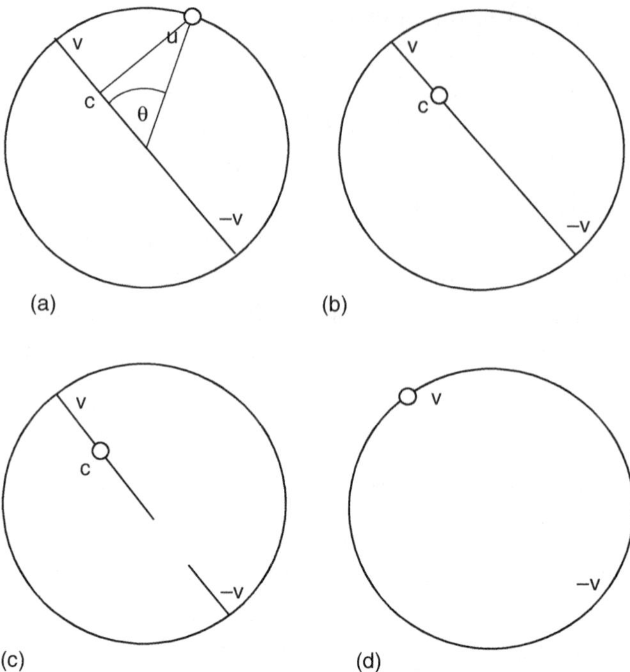

Figure 10.4. The four stages of a measurement in the elastic band model.

isomorphic to the Bloch sphere). Given a direction v, the pure states associated to this direction will be given by

$$
W(v) = \begin{pmatrix} \cos^2\dfrac{\theta}{2} & e^{-i\varphi}\cos\dfrac{\theta}{2}\sin\dfrac{\theta}{2} \\[2ex] e^{i\varphi}\cos\dfrac{\theta}{2}\sin\dfrac{\theta}{2} & \sin^2\dfrac{\theta}{2} \end{pmatrix} \tag{20}
$$

and

$$
W(-v) = \begin{pmatrix} \sin^2\dfrac{\theta}{2} & -e^{-i\varphi}\cos\dfrac{\theta}{2}\sin\dfrac{\theta}{2} \\[2ex] -e^{i\varphi}\cos\dfrac{\theta}{2}\sin\dfrac{\theta}{2} & \cos^2\dfrac{\theta}{2} \end{pmatrix} \tag{21}
$$

for the orthogonal direction. Thus, a general mixed state can be written as:

$$
W(x) = aW(v) + bW(-v) \tag{22}
$$

provided that $X = av + b(-v)$. In turn, entanglement can be simulated in this model by placing a rigid rod between two balls in the elastic band model. In this way, the states of the balls can be expressed in a form mathematically equivalent to quantum entangled states (see Aerts 1998 for details).

Recently, a striking example of bouncing droplets of a non-coalescent liquid has been introduced (Couder et al. 2005; Couder and Fort 2006). The droplets generate waves as they bounce, and interact with their own waves. These waves also interact with the environment (and other bouncing droplets), generating a complex dynamics that resembles that of particles guided by pilot waves. It has been experimentally demonstrated that bouncing droplets can be used to reproduce interference phenomena, tunneling, and quantum random walk. Of course, the existence of these models does not prove the existence of pilot waves in the real world. But it does show how an atomic-scale pilot wave might work. And as it might be expected, these bouncing droplets gave birth to a new interpretation of quantum mechanics (Brady and Anderson 2014).

The most useful physical devices to study information protocols are those based on electromagnetic fields. In fact, it has been proved that the behavior of quantum computers and quantum information protocols can be simulated by means of classical optical waves. A possible strategy to do this for a system of n qubits is to consider the profile of the classical electric-field amplitude in a laser beam as the analog of the probability amplitude of a quantum state. Although this simulation is inefficient, since it requires a number of classical resources that scales exponentially with the number of quantum bits being simulated, this kind of states of the electromagnetic field has been used in several works to study different aspects of quantum information theory. For example, Puentes and colleagues (2004) introduce an optical simulation of quantum algorithms by using programmable liquid-crystal displays. Cerf and colleagues (1998) show that a nontrivial quantum computing optical device can easily be constructed if the number of component qubits is not too large. In turn, Spreeuw (1998) presents a classical analog of entanglement, but proves that it cannot replace Einstein-Podolsky-Rosen type experiments, nor can it be used to build a quantum computer (see also Spreeuw 2001). Kwiat and colleagues (2000) demonstrate that the essential operations of a quantum computer can be accomplished by means of solely optical elements, with different polarization or spatial modes representing the individual qubits (see also Bhattacharya et al. 2002). These are only some examples of how optical simulation has been applied to the computational context (for a simulation of a Hadamard gate, see Francisco et al. 2006a; a simulation of quantum teleportation protocol is presented in Francisco and Ledesma 2008; a simulation of a quantum walk is presented in Francisco et al. 2006b).

Let us see how this works by following a concrete example (we follow Aiello et al. 2015, and refer the reader to this reference for details). A classical electric field can be described mathematically as follows:

$$E(r, z, t) = 2\mathrm{Re}[U(r, z)\exp(ik(z - ct))] \tag{23}$$

where $U(r, z)$ is a complex number called the *analytic signal*, depending on spatial coordinates r (radial) and z (longitudinal), and time t. We can write

$$U(r, z) = \boldsymbol{a}_1 b_1(r, z) + \boldsymbol{a}_2 b_2(r, z) \tag{24}$$

where $\boldsymbol{a}_1$ and $\boldsymbol{a}_2$ are two orthogonal vectors and $b_1(r, z)$ and $b_2(r, z)$ are two solutions of the paraxial equation. These functions satisfy $\int b_1^*(r, z) b_2(r, z) dx dy = 0$. One particular form of interest is that of *polarization-position classical entanglement*. In order to obtain this case, the field is prepared in the following state:

$$U(r, z) = \boldsymbol{e}_H U_{up}(r, z) + \boldsymbol{e}_V U_{down}(r, z) \tag{25}$$

as the simulation of:

$$|\Psi\rangle = |H\rangle|up\rangle + |V\rangle|down\rangle \tag{26}$$

Another possibility is using only spatial modes, leading to *position-position classical entanglement* as follows:

$$\begin{aligned}U(\rho, z) = A_{00}U(x + a, y - a) + A_{01}U(x - a, y - a) + \\ A_{10}U(x + a, y + a) + A_{11}U(x - a, y + a)\end{aligned} \tag{27}$$

as the simulator of two qubits. The wave functions satisfy:

$$U(x + (-1)^j a, y - (-1)^i a) = rec(y - (-1)^i ab)rec(x + (b - 1)^j a \tag{28}$$

The last form is that of *polarization-spatial classical entanglement*, for which a state is prepared as:

$$U(\rho, z) = \boldsymbol{e}_H U_{10}(\rho, z) + \boldsymbol{e}_V U_{01}(\rho, z) \tag{29}$$

as a simulation of

$$|\Psi\rangle = |H\rangle|+\rangle + |V\rangle|-\rangle \tag{30}$$

where $U_{10}(\rho, z)U_{01}(\rho, z) \neq 0$ but $\int U_{10}(\rho, z)U_{01}(\rho, z)dx dy = 0$. Notice that, in this case, the wavefunctions are orthogonal but the overlap is non null (as in the previous cases).

3.4 Quantum Information Processing and Simulators

The discussion of Section 3.3 shows that it is possible in principle to perform many quantum information protocols by using classical systems. This is due to the fact that it is possible to simulate the mathematical formalism of quantum mechanics (and other non-classical theories as well) with an essentially classical ontology. As a radical example, consider Bohmian mechanics. According to the ontology of this theory, the world is made of point particles with definite trajectories (which evolve in a deterministic way), but interacting via a highly non-local potential, which is responsible for the complementarity observed in the laboratory. Or to put it in different terms: in a world strictly governed by Newtonian mechanics, non-local correlations could be reproduced! Newton could have conceived and developed a non-classical computer if the world would have been really Newtonian, due to the existence of rigid rods and instantaneous gravitational forces. In other words, if non-local interactions are allowed, it is possible to use an essentially classical ontology and simulate complementarity and non-classical information protocols.

The key observation here is that the same mathematical formalism, with essentially the same empirical content from the operational point of view, can be supported by very different ontologies. This means that classical systems are ontological alternatives for the realization of the mathematical formalism of quantum information theory. Then, we see that at the metaphysical level we can perform quantum information processing without the need of quantum mechanical models.

As we have seen in Section 2, one can consider an abstract mathematical formalism for a generalized information theory. This formalism does not have a priori any essential connection with any physical system. Of course, one may endow the elements of the formalism with an operational significance: this is the content of the operational approach to information theory. But the ontology of this formal construction still remains undefined. This metaphysical underdetermination can be stated as follows: the formalism and the operational empirical content of a given theory are not sufficient to determine its ontological reference. A clear example of this is quantum mechanics itself: there are many interpretations, with radically different ontologies, which are coincident with regard to their operational content. These observations preclude us to assign quantum systems as the only possible instances of the quantum information formalism.

What is the difference between classical and quantum then? The main difference seems to rely on effectiveness with regard to algorithmic complexity: tasks exist that can be solved in polynomial time using quantum systems, but take exponential time if we use classical simulations. And this seems to be a rule: all the examples show the exponential growth in resources when classical simulators are used.

Summarizing: both classical and quantum systems can be used as models of the quantum formalism described in Section 2, but quantum systems are faster. Where does this velocity come from? Is effectiveness a suitable criterion to distinguish the "classicality" or "quantumness" of two given interpretations of an information theory?

At this point it seems that the formalisms presented in Section 2 are not sufficient by themselves to decide about its ontological reference. In order to distinguish between classical and quantum by means of information formalisms, we need to consider new formal tools, such as computational complexity. If we want to single out quantum systems, we need to add the consideration of the energetic resources consumed and the duration of the operational processes involved to the formal/operational apparatus. But again, there should be no confusion about what we mean by "classical" here: limitations arise in the case of relativistic theories, endowed with essentially local features. However, the entire formal apparatus of quantum information theory can be reproduced in terms of Bohmian mechanics, or even of Newtonian mechanics with sufficiently non-local interactions.

4 Conclusions

In classical theories, we deal with event spaces forming Boolean algebras; in the quantum case and more general models, we deal with contextual (non-Boolean) event spaces. As probabilities lie at the heart of the information theory in Shannon's version, it is not surprising to find contextual versions of this theory in more general probabilistic models. In particular, quantum information theory is the mathematical version of the information theory based on quantum formalism. One may conjecture that there is an information theory for each family of probabilistic models (Boolean algebras, Type I, II and III factors, and so on; see Holik 2014). But then the question arises: what does it lie behind those mathematical formalisms? Which are the ontological correlates of these probabilistic models (including the quantum one)?

Can we speak consistently of *quantum* information? In this chapter, we have presented some arguments and examples that might help to answer the question; but much work must still be done in this direction in order to formulate a cogent ontological reference for quantum information theory. An important step in this direction is to clarify the scope and limitations of classical models of quantum information. We have seen that, in many senses, it is possible to conceive models that embrace a classical ontology, but from the formal point of view exhibit quantum features; these features make them acceptable as models of quantum information formalism. The existence of these models seems to point in the direction that, physically, there is no quantum commitment in the formalism of quantum information theory; quite on the contrary, dealing with concrete quantum systems is not necessary at all in order to cope with the more important features of quantum information theory. We can conclude that, at the ontological level, the formalism of quantum information does not require actual quantum systems.

Nevertheless, these considerations do not imply that the technological developments based on quantum systems are trivial; quite on the contrary, they constitute the intended interpretation of the formalism and are, up to now, the only known candidates practically available for realizing the formalism in an *efficient* way, offering a host of promising technologies. What is the difference between classical and quantum, then? It is true that ontologically classical theories endowed with strong non-local features, such as strict Newtonian mechanics, can mimic several quantum protocols. However, the world is not Newtonian, but relativistic and, therefore, it is essentially local at the classical level. In this world, for any classical simulator there is a drawback in the exponential increase of the resources required for the simulation. The relatively extensive literature on the subject shows that the classical models for quantum information protocols cannot replace Einstein-Podolsky-Rosen-type experiments, nor can it be used to build efficient quantum computers.

Summing up, the results presented in this chapter show that, if we want to understand the differences between classical information and quantum information, we need to examine the level of computational complexity. Very likely, the origin of this peculiar feature of quantum mechanics relies on irreconcilable ontological differences with classical physics, but this will be the subject of another article.

Acknowledgments

Federico Holik and Gustavo Martín Bosyk acknowledge CONICET (Argentina) and Universidad Nacional de La Plata (Argentina). Federico also acknowledges the support of a Large Grant of the Foundational Questions Institute (FQXi).

References

Aaronson, S. and Arkhipov, A. (2011). "The Computational Complexity of Linear Optics." Pp. 333–342 in *Proceedings of the 43rd Annual ACM Symposium on Theory of Computing*. New York: Association for Computing Machinery.

Aerts, D. (1998). "The Hidden Measurement Formalism: What can Be Explained and Where Quantum Paradoxes Remain." *International Journal of Theoretical Physics*, **37**: 291–304.

Aiello, A., Töppell, F., Marquardt, C., Giacobino, E., and Leuchs, G. (2015). "Quantum-Like Non-separable Structures in Optical Beams." *New Journal of Physics*, **17**: 043024.

Amaral, B., Terra Cunha, M., and Cabello, A. (2014). "Exclusivity Principle Forbids Sets of Correlations Larger than the Quantum Set." *PhysicsReview A*, **89**: 030101.

Barnum, H., Barrett, J., Clark, L., Leifer, M., Spekkens, R., Stepanik, N., Wilce, A., and Wilke, R. (2010). "Entropy and Information Causality in General Probabilistic Theories." *New Journal of Physics*, **12**: 033024.

Beltrametti, E. and Cassinelli, G. (1981). *The Logic of Quantum Mechanics. Encyclopedia of Mathematics and its Applications, Volume 15*. Reading MA: Addison-Wesley.

Bennett, C. (1995). "Quantum Information and Computation." *Physics Today*, **48**: 24–30.

Bhattacharya, N., Van Linbiden, H., Van den Heuvell, L., and Spreeuw, R. (2002). "Implementation of Quantum Search Algorithm Using Classical Fourier Optics." *Physics Review Letters*, **88**: 137901.

Brady, R. and Anderson, R. (2014). "Why Bouncing Droplets Are a Pretty Good Model of Quantum Mechanics." arXiv:1401.4356v1 [quant-ph].

Brunner, N., Cavalcanti, D., Pironio, S., Scarani, V., and Wehner, S. (2014). "Bell Nonlocality." *Reviews of Modern Physics*, **86**: 419–478.

Bub, J. (2007). "Quantum Computation from a Quantum Logical Perspective." *Quantum Information & Computation*, **7**: 281–296.

Cabello, A. (2014). "Exclusivity Principle and the Quantum Bound of the Bell Inequality." *Physics Review A*, **90**: 062125.

Carolan, J., Harrold, C., Sparrow, C., Martín-López, E., and Russell, N. J. (2015). "Universal Linear Optics." *Science*, **349**: 711–716.

Cerf, N., Adami, C., and Kwiat, P. (1998). "Optical Simulation of Quantum Logic." *Physical Review A*, **57**: 1477–1480.

Clifton, R., Bub, J., and Halvorson, H. (2003). "Characterizing Quantum Theory in Terms of Information-Theoretic Constraints." *Foundations of Physics*, **33**: 1561–1591.

Couder, Y. and Fort, E. (2006). "Single-Particle Diffraction and Interference at a Macroscopic Scale." *Physical Review Letters*, **97**: 154101.

Couder, Y., Fort, E., Gautier, C., and Boudaoud, A. (2005). "From Bouncing to Floating: Noncoalescence of Drops on a Fluid Bath." *Physical Review Letters*, **94**: 177801.

Cox, R. T. (1946). "Probability, Frequency and Reasonable Expectation." *American Journal of Physics*, **14**: 1–13.

Cox, R. T. (1961). *The Algebra of Probable Inference*. Baltimore, MD: Johns Hopkins University Press.

de Barros, J. A. and Oas, G. (2014). "Negative Probabilities and Counter-factual Reasoning in Quantum Cognition." *Physica Scripta*, **163**: 014008.

de Barros, J. and Oas, G. (2015). "Some Examples of Contextuality in Physics: Implications to Quantum Cognition." Pp. 153–184 in E. Dzhafarov, S. Jordan, R. Zhang, and V. Cervantes (eds.), *Contextuality from Quantum Physics to Psychology*. Singapore: World Scientific.

De Finetti, B. (1937). "La Prévision: ses Lois Logiques, ses Sources Subjectives." *Annales de l'Institut Henri Poincaré*. Reprinted as "Foresight: Its Logical Laws, Its Subjective Sources." In H. E. Kyburg and H. E. Smokler (eds.), *Studies in Subjective Probability*, 1964. New York: Wiley.

Deutsch, D. and Ekert, A. (2000). "Introduction to Quantum Computation." Pp. 92–131 in D. Boumeester, A. Ekert and A. Zeilinger (eds.), *The Physics of Quantum Information*. Berlin-Heidelberg: Springer.

Di Vicenzo, D. (1995). "Quantum Computation." *Science*, **270**: 255–261.

Everett, H. (1957). "Relative State Formulation of Quantum Mechanics." *Reviews of Modern Physics*, **29**: 454–462.

Feynman, R. (1951). "The Concept of Probability in Quantum Mechanics." Pp. 533–541 in J. Neyman (ed.), *Proceedings of the Second Berkeley Symposium on Mathematical Statistics and Probability*. Berkeley: University of California Press.

Francisco, D., Iemmi, C., Paz, J. P., and Ledesma, S. (2006a). "Optical Simulation of the Quantum Hadamard Operator." *Optics Communications*, **268**: 340–345.

Francisco, D., Iemmi, C., Paz, J. P., and Ledesma, S. (2006b). "Simulating a Quantum Walk with Classical Optics." *Physical Review A*, **74**: 052327.

Francisco, D. and Ledesma, S. (2008). "Classical Optics Analogy of Quantum Teleportation." *Journal of the Optical Society of America B*, **25**: 383–390.

Fuchs, C. and Schack, R. (2013). "Quantum-Bayesian Coherence." *Reviews of Modern Physics*, **85**: 1693–1715.

Gleason, A. (1957). "Measures on the Closed Subspaces of a Hilbert Space." *Journal of Mathematics and Mechanics*, **6**: 885–893.

Goldin, M., Francisco, D., and Ledesma, S. (2010). "Simulating Bell Inequality Violations with Classical Optics Encoded Qubits." *Journal of the Optical Society of America B*, **27**: 779–786.

Gühne, O., Kleinmann, M., Cabello, A., Larsson, J., Kirchmair, G., Zahringer, F., Gerritsma, R., and Roos, C. (2010). "Compatibility and Noncontextuality for Sequential Measurements." *Physical Review A*, **81**: 022121.

Holik, F. (2014). "Logic, Geometry and Probability Theory." *SOP Transactions on Theoretical Physics*, **1**: 128–137.

Holik, F., Bosyk, G. M., and Bellomo, G. (2015). "Quantum Information as a Non-Kolmogorovian Generalization of Shannon's Theory." *Entropy*, **17**: 7349–7373.

Holik, F. and Plastino, A. (2014). "Quantum Mechanics: A New Turn in Probability Theory." Pp. 399–414 in Z. Ezziane (ed.), *Contemporary Research in Quantum Systems*. New York: Nova Science.

Holik, F., Plastino, A., and Sáenz, M. (2014). "A Discussion on the Origin of Quantum Probabilities." *Annals of Physics*, **340**: 293–310.

Holik, F., Plastino, A., and Sáenz, M. (2016). "Natural Information Measures for Contextual Probabilistic Models." *Quantum Information & Computation*, **16**: 115–133.

Jozsa, R. (2000). "Quantum Algorithms." Pp. 104–125 in D. Boumeester, A. Ekert, and A. Zeilinger (eds.), *The Physics of Quantum Information*. Berlin-Heidelberg: Springer.

Kolmogorov, A. (1933). *Foundations of Probability Theory*. Berlin: Julius Springer.

Kwiat, P., Mitchell, J., Schwindt, P., and White, A. (2000). "Grover's Search Algorithm: An Optical Approach." *Journal of Modern Optics*, **47**: 257–266.

Landauer, R. (1996). "The Physical Nature of Information." *Physics Letters A*, **217**: 188–193.

Lombardi, O., Holik, F., and Vanni, L. (2016a). "What Is Quantum Information?" *Studies in History and Philosophy of Modern Physics*, **56**: 17–26.

Lombardi, O., Holik, F., and Vanni, L. (2016b). "What Is Shannon Information?" *Synthese*, **193**: 1983–2012.

Maassen, H. and Uffink, J. (1988). "Generalized Entropic Uncertainty Relations." *Physical Review Letters*, **60**: 1103–1106.

Man'ko, M. A., Man'ko, V. I., and Vilela Mendes, R. (2001). "Quantum Computation by Quantumlike Systems." *Physics Letters A*, **288**: 132–138.

Nielsen, M. and Chuang, I. (2010). *Quantum Computation and Quantum Information*. Cambridge: Cambridge University Press.

Oas, G. and de Barros, J. A. (2015). "Principles Defining Quantum Mechanics." Pp. 335–366 in E. Dzhafarov, S. Jordan, R. Zhang, and V. Cervantes (eds.), *Contextuality from Quantum Physics to Psychology*. Singapore: World Scientific.

Popescu, S. (1994). "Bell's Inequalities versus Teleportation: What Is Nonlocality?" *Physical Review Letters*, **72**: 797–799.

Popescu, S. (2014). "Nonlocality Beyond Quantum Mechanics." *Nature Physics*, **10**: 264–270.

Puentes, G., La Mela, C., Ledesma, S., Iemmi, C., Paz, J. P., and Saraceno, M. (2004). "Optical Simulation of Quantum Algorithms Using Programmable Liquid-Crystal Displays." *Physical Review A*, **69**: 042319.

Qian, X.-F., Little, B., Howell, J. C., and Eberly, J. H. (2015). "Shifting the Quantum-Classical Boundary: Theory and Experiment for Statistically Classical Optical Fields." *Optica*, **2**: 611–615.

Renes, J. (2013). "The Physics of Quantum Information: Complementarity, Uncertainty and Entanglement." *International Journal of Quantum Information*, **11**: 1330002.

Schumacher, B. (1995). "Quantum Coding." *Physical Review A*, **51**: 2738–2747.

Shannon, C. (1948). "The Mathematical Theory of Communication." *Bell System Technical Journal*, **27**: 379–423, 623–656.

Short, A. and Wehner, S. (2010). "Entropy in General Physical Theories." *New Journal of Physics*, **12**: 033023.

Spreeuw, R. (1998). "A Classical Analogy of Entanglement." *Foundations of Physics*, **28**: 361–374.

Spreeuw, R. (2001). "Classical Wave-Optics Analogy of Quantum-Information Processing." *Physical Review A*, **63**: 062302.

Steane, A. (1998). "Quantum Computing." *Reports on Progress in Physics*, **61**: 117–173.

11

On the Relative Character of Quantum Correlations

ÁNGEL LUIS PLASTINO, GUIDO BELLOMO, AND ÁNGEL RICARDO PLASTINO

1 Introduction

Nowadays, no one should doubt that quantum entanglement is (at least, one of) the characteristic traits of quantum mechanics. Furthermore, quantum information science is strongly influenced by the advantages provided by quantum entanglement, as the latter can be regarded as a genuine resource for an efficient (better-than-classical) processing of information in certain computational tasks. The very basic notion of correlation is fundamental for the conceptualization of the physical communicational problem and, as a consequence, for the notion of information itself. Thus, the question about the nature of quantum information is close (and unavoidably linked) to the question about the nature of quantum correlations.

In this chapter, our aim is to discuss how to assess the concepts of quantum correlations and quantumness – where "quantumness" is used in a more generic way than the term "quantum correlations," not referring to an amount of quantum information, but to its non-classical character. We look at things from a perspective general enough so as to enable the development of a formal framework that is suitable for the explanation of any nonrelativistic quantum model. Also, we emphasize the clear distinction between locality and quantumness, two notions usually subsumed under the entanglement concept. These goals are tackled by revisiting these topics from a generalized perspective that focuses on the relational aspect of the term "correlation" with respect to the states and observables involved.

We begin by introducing, in Section 2, the formal concepts necessary for the present discussion, with particular emphasis on the relations between quantum observables, quantum states, and the way in which a system is partitioned into subsystems. Also, we introduce the notions of entanglement and discord, two kinds of well-known quantum correlations that we propose to review and to generalize. In Section 3, we present our main discussion, regarding what is called "the relative nature of quantum correlations." We start from the fact that the correlation concept is inherently relative to the non-unique partition of a system into subsystems.

Furthermore, we favor a subsystem-independent view of the quantum correlations that provides what we call "a second stage of relativeness." Such analysis enables generalized definitions for entanglement and discord. In Sections 4 and 5, we discuss some issues involving this relative character of quantum correlations and consider the obstacles that make it difficult to achieve a unified treatment for them, including the relativity/ambiguity debate, the status of properties such as monogamy or convexity, and the emergence of the classical world from the quantum world. Finally, in Section 6, we review the main points of this chapter, which leads to some concluding remarks.

2 Quantum Systems and Correlations

Before tackling the discussion of the peculiarities of quantum correlations, we should establish some basic concepts. How do we describe a quantum system? What is the role of observables and states? When do we speak about correlations? Whose correlations?

Our presentation relies on the algebraic approach to quantum mechanics. We give here a brief introduction, referring the readers to the books by Bratteli and Robinson (2012) and Haag (2012) – and references therein – for a deeper analysis. One starts by defining an *observable algebra* $\Omega(\mathcal{O})$, which is the algebra spanned by a certain set $\mathcal{O}$ of relevant observables O represented by self-adjoint operators mapping a suitable Hilbert space, $\mathcal{H}$, onto itself. As an example, consider a single spin one-half particle for which we should choose $\Omega(\mathcal{O})$ to be the algebra of the Lie group SU(2), spanned by the identity operator, $\mathbb{I}$, and the three Pauli matrices, σ_i with $i = x, y, z$, acting on $\mathbb{C}^2$. Otherwise, for a single particle in a d-dimensional Euclidean space, $\mathbb{R}^d$, $\Omega(\mathcal{O})$ may be the algebra $\mathcal{B}\left(L^2_{\mathbb{C}}(\mathbb{R}^d)\right)$ spanned by the self-adjoint operators acting on the Hilbert space of square integrable complex-valued functions, $L^2_{\mathbb{C}}(\mathbb{R}^d)$. Each algebra $\Omega(\mathcal{O})$ identifies a quantum system. A *state* of the system is a prescription of the expectation values of the observables, and it is formalized as an expectation value functional from the observables to the unit interval, $\omega : \Omega(\mathcal{O}) \to [0, 1]$. The state has (when exists) an associated density operator, ρ_ω (with $\rho_\omega \geq 0$ and $\mathrm{Tr}(\rho_\omega) = 1$), acting on the same Hilbert space, $\mathcal{H}$. As stated previously, we have a law linking the observables and its expectation values with the state, namely $\omega(O) = \mathrm{Tr}(O\rho_\omega)$ for all $O \in \mathcal{O}$. This quantity, $\omega(O)$, which is called the "expectation value," gives the (probabilistic) expected value if one measures the observable O when the system is in the state ω. This is, essentially, the Born rule (1926) extended to a mixed-states setting. It is important to stress that $\omega(O)$ is not the most probable outcome for that measurement (indeed, $\omega(O)$ is an impossible outcome if it does not coincide with an eigenstate of O), but the probabilistic average for its outcomes.

An important class of states is the one constituted by the states that cannot be written as a non-trivial convex combination of other states, i.e., as $\omega = a\omega_1 + (1 - a)\omega_2$, where $0 < a < 1$ and ω_1, ω_2 are distinct states on $\Omega(\mathcal{O})$. Those particular states are called "pure." Every non-pure state is a *mixed state*. In principle, pure states imply maximal knowledge about our system because they are the eigenstates of a global observable (implying that there is an observable that can be measured with no uncertainty). Geometrically, pure states are extremal states from the convex set of quantum states (see Bengtsson and Zyczkowski 2006 for a complete description of the quantum states from a geometrical point of view).

Among observables, there is also a special class: some elements $E \in \Omega(\mathcal{O})$ that are not only self-adjoint, but also idempotent: $E^2 = E$. In that case, we say that E is a *projection*. Projections are associated to "Yes-No" questions such that, for a state ω with density operator ρ_ω, $\omega(E) = \mathrm{Tr}(E\rho_\omega)$ is the probability of a "Yes" outcome if one measures E. From a logic-theoretical viewpoint, the set of projections, $\mathcal{P}(\Omega)$, has the structure of a non-abelian (or non-commutative) ortho-complemented lattice (see, e.g., Rédei 2013). Moreover, given the algebra $\Omega(\mathcal{O})$ and a state ω, there is a countably additive probability measure on $\mathcal{P}(\Omega)$ associated to that state and, due to the generalized Gleason theorem (see Hamhalter 2003 for details), this probability measure has a unique extension to a state on $\Omega(\mathcal{O})$. The dynamical evolution of the system is determined by its Hamiltonian, represented by another operator over the same Hilbert space, through the Liouville-von Neumann equation. If the system is open, the evolution follows a particular non-unitary law that depends on the interactions between the system and its environment. For the most of our discussion, the dynamics is not important and thus our main ingredients to characterize a system are the algebra of relevant observables and the states.

Furthermore, we are interested in discussing correlations, i.e., statistical relations between different properties. In particular, we are interested in correlations between properties that are associated to different systems (or, analogously, between the subsystems of a given system). In either case, as we already stated, different (sub) systems are characterized by their own relevant observables and their corresponding algebras. Although quantum subsystems are usually thought in terms of spatially localized degrees of freedom in space-like separated regions, none of these is necessary for the usual notions of quantum correlations to be suitably defined. On the other hand, some kind of *independence between subsystems* can be formalized under the present algebraic treatment: mutual commutativity of subsystem algebras is a good starting point, allowing for the possibility to simultaneously measure observables of different subsystems. Herefrom, we adopt this notion of independence as a condition before asking questions about correlations between different observables/subsystems. There are many other alternative ways of defining subsystems' independence. A rather complete survey has been given by Rédei (2013).

Finally, we should mention some concerns about the non-abelian character of the algebras of quantum observables. Two non-commutative observables cannot be simultaneously measured, in the sense that the order in their measurement process leads to different results. Moreover, the measurement of any observable over a state that has not been prepared in one of its eigenstates disturbs the state. Thus, non-commutativity of the observables turns out to be a key concept underlying the first notions of discord-like correlations (see Subsection 2.2). As quantum states and observables usually live in non-abelian algebras, an operation over the state can disturb it, leading to a post-measurement state that is different from the original one. Any observable, being represented by a self-adjoint operator that maps a Hilbert space into itself, has associated eigenvectors that form an orthonormal basis spanning the state space in which that observable exists. Thus, any operator can be written as a superposition of the projections linked to those eigenvectors. For example, assume that for a system in a state given by ρ_ω, we measure an observable O such that E_k is the projection into its k-th eigenspace (the one spanned by the k-th corresponding eigenvector), with o_k eigenvalue. The measurement will yield any of the results o_k with probability $\Pr(o_k; \omega) = \mathrm{Tr}(E_k \rho_\omega)$. If the result of the measurement is o_k, the post-measurement state is the projection of ρ_ω into the k-th eigenspace, correspondingly normalized, i.e., $\rho_\omega^{(k)} = (E_k \rho_\omega E_k)/\mathrm{Tr}(E_k \rho_\omega)$. That kind of measurement is called "selective," with E_k being the corresponding measurement operator. If, on the contrary, the measurement remains unread (alternatively, one says that a non-selective measurement is performed) the post-measurement state is, according to the Lüders rule, the ensemble $\rho'_{\omega'} = \sum_k E_k \rho_\omega E_k = \sum_k \Pr(o_k; \omega)\rho_\omega^{(k)}$.

2.1 Subsystems and Correlations

When we deal with classical probability theory, the situation is as follows. Two random variables, A and B, are said to be correlated if their joint probability distribution, $P(a, b)$, cannot be expressed as a simple product between the marginal probabilities $P(a)$ and $P(b)$, for all $a \in A$ and $b \in B$. That is, if $P(a, b)$ represents the state of certain classical system then the properties associated with A and B are correlated if $P(a, b) \neq P(a)P(b)$. When these variables A and B correspond to two different subsystems, we say that both subsystems are correlated. In quantum mechanics, each subsystem is characterized by 1) its own observables $A \in \mathcal{A}$ and $B \in \mathcal{B}$, and 2) the corresponding (mutually commutative) algebras $\Omega(\mathcal{A})$ and $\Omega(\mathcal{B})$, acting on a common Hilbert space. The joint state ω lives in $\Omega(\mathcal{A}) \vee \Omega(\mathcal{B})$, the minimal algebra generated by $\Omega(\mathcal{A})$ and $\Omega(\mathcal{B})$. Then, the non-correlation condition reads (see Definition 1): $\omega(AB) \neq \omega(A)\omega(B)$. It is straightforward to see that *the notion of correlation depends on both the observables and the state*. Thus, we should rigorously say that our subsystems are in a correlated state with respect to the observables A and B. It is clear, then, that correlations refer to some particular observables of a system in a certain state. We reproduce here the definition of product states due to Earman (2014):

Definition 1 (Non-correlation). Let $\Omega(\mathcal{A})$ and $\Omega(\mathcal{B})$ be two mutually commutative algebras of observables. A state ω over $\Omega(\mathcal{A}) \vee \Omega(\mathcal{B})$ is non-correlated with respect to $\Omega(\mathcal{A})$ and $\Omega(\mathcal{B})$ if $\omega(AB) = \omega(A)\omega(B)$, for all $A \in \mathcal{A}$ and $B \in \mathcal{B}$. Equivalently, we say that ω is a *product* state with respect to those algebras.

This notion of correlations is borrowed from the traditional definition in classical probabilistic theories. What we say here is that two subsystems are in a correlated state if they share any pair of correlated variables. The main difference resides in the mutual-commutativity condition that one must impose given the non-abelian nature of the algebra of quantum observables. To understand the richness of the structure of correlations in quantum theory, distinction of correlated versus product states is not enough. Genuine quantum correlations (as entanglement and discord) attempt to capture that richness.

2.2 Quantum Entanglement and Discord

Correlations tell us how different properties of a system are interrelated. In the quantum case, correlations predictions concern the expectation values of different observables that are simultaneously measured. In general, two arbitrary observables do not commute and, even more, they do not commute with the state of our system. Thus, we have the following consequences: (a) measuring properties disturbs the state of the system, (b) the order in which we measure them determines the results. Both facts rely on the non-abelian (non-commutative) character of the algebra of observables, and lead to the non-classical correlations known as *entanglement* or *discord*. In classical mechanics, on the contrary, the algebra of observables is abelian (commutative) and so there is no place for entanglement or discord.

There are many definitions of entanglement and discord-like quantities in the literature. We adopt the following definitions, which are consistent with the ones frequently encountered nowadays (see the recent work by Earman 2014). Let us start by defining what entanglement is.

Definition 2 (Entanglement). Let $\Omega(\mathcal{A})$ and $\Omega(\mathcal{B})$ be two mutually commutative algebras of observables. A state ω over $\Omega(\mathcal{A}) \vee \Omega(\mathcal{B})$ is non-entangled with respect to $\Omega(\mathcal{A})$ and $\Omega(\mathcal{B})$ if $\omega(AB)$ can be written as a convex sum of product states, for all $A \in \mathcal{A}$ and $B \in \mathcal{B}$. Else, we say that ω is an entangled state with respect to those algebras.

The basis for this definition can be traced to the works by Raggio (1988) and Werner (1989). There are other notions of entanglement that consider the violation of Bell-type inequalities or the impossibility of a hidden variable model. But Definition 2, in terms of decomposability of the state operator, is the most used one today. It is noteworthy the observation by Earman (2014) regarding the connection between this notion of entanglement and the non-abelian character of the algebras of observables, based on the result, due to Raggio (1988): *entanglement is possible if and only if one of the subsystem algebras is non-abelian.*

As noted before, the notion of discord is also closely linked to this non-abelian nature of the algebras of quantum observables. The first definition of discord was proposed by Ollivier and Zurek (2001) and, independently, by Henderson and Vedral (2001). From a purely informational motivation we should emphasize that a classical state must be such that the information content of correlations can be locally accessed without disturbing the joint state. This idea, applied to the bipartite setting, is formalized by computing the difference between the mutual information of the state and the mutual information of the post-local-measurement state. For a bipartite system, the measurements can be performed locally on the first part or on the second part (or simultaneously on both). Hence, the following definition is proposed:

Definition 3 (Discord). Let $\Omega(\mathcal{A})$ and $\Omega(\mathcal{B})$ be two mutually commutative algebras of observables. A state ω over $\Omega(\mathcal{A}) \vee \Omega(\mathcal{B})$ is non-discordant with respect to $\Omega(\mathcal{A})$ if there is a local complete projective measurement, given by $\{E_k^A\}$, such that leaves the state invariant, i.e., $\rho_\omega = \sum_k E_k^A \rho_\omega E_k^A$. Equivalently, we say that ω is $\mathcal{A}$-non-discordant.

To define a complete measurement, the set $\{E_k^A\}$ must be a minimal resolution of the identity, that is, $\sum_k E_k^A = \mathbb{I}^A$ with E_k^A being one-dimensional projectors. Interchanging the roles of $\mathcal{A}$ and $\mathcal{B}$ provides the definition for $\mathcal{B}$-non-discordant states. In addition, we define the $\mathcal{AB}$-non-discordant states considering complete bi-local measurements given by $\{E_k^A \otimes E_l^B\}$. If we demand non-discordance with respect to every (including non-local) complete projective measurement and for any state, we are demanding that the observables must commute (Earman 2014). Thus, for a system to be not disturbed under any complete projective measurement, the algebra of observables must be abelian.

Nowadays, many measures of quantum correlations exist, most of them inspired by an information-theoretic approach to quantum mechanics. These novel attempts try to connect entanglement with discord, non-locality, coherence, complementarity, contextuality, or geometry of the state-space. We focus on entanglement and discord as they are two of the most relevant measures in the current specialized literature regarding quantum information and quantum foundations. For more details, we refer the reader to the excellent reviews by Horodecki and colleagues (2009) and Modi and colleagues (2012).

2.3 Information-Theoretic Quantifiers of Quantum Correlations

Most of the concepts discussed so far have their counterparts in terms of information-theoretic properties of the states and observables. We are just going to review some basic concepts that the reader may find in the standard literature of quantum information (e.g., Nielsen and Chuang 2010). Let us remind that, for a given state ω with spectrum $\{\lambda_i, |\omega_i\rangle\}$, one defines its von Neumann entropy as $S(\omega) = -\sum_i \lambda_i \log \lambda_i = -\mathrm{Tr}(\rho_\omega \log \rho_\omega)$. It is noteworthy that, if one defines a measurement entropy as $S(\omega; \{p_i, E_i\}) = -\sum_i p_i \log p_i$, with p_i the probability associated to the measurement operator E_i, the minimum measurement entropy turns out to be the von Neumann entropy of ω. In that sense, the measurement in the eigenbasis of the states is optimal. For pure states, states of maximal knowledge, the entropy is null. For the maximally mixed state, meaning total ignorance, the entropy achieves its maximum value. Thus, von Neumann entropy quantifies, in some way, the mixedness/non-purity or the information content of the state.

Consideration of bipartite scenarios, with ω over $\Omega(\mathcal{A}) \vee \Omega(\mathcal{B})$, yields a couple of notable results based on the subadditivity of S: $S(AB) \leq S(A) + S(B)$, where $S(AB)$ stands for $S(\omega(AB))$ and so on. Subadditivity shows us that, even when we have total knowledge of the joint state, i.e., $S(AB) = 0$, we have incomplete knowledge of the states of its parts. The extremal case is the one of maximally entangled pure states, when both $S(A)$ and $S(B)$ are maximal while $S(AB)$ is zero. Hence, for pure states, $S(A) = S(B)$ is a faithful indicator of entanglement. Classically, for a probability vector $\vec{p} = (p_i)$, one defines the Shannon entropy $H(\vec{p}) = -\sum_i p_i \log p_i$, which, besides compelling also to a subadditivity relation, obeys the relation $H(AB) \geq \max(H(A), H(B))$. In particular, if $H(AB) = 0$, both the joint states and its marginals are pure. Thus, classically, the knowledge of the joint state is always lower bounded by the knowledge of the states of its marginals. An analogous relation for quantum states, namely that $S(AB) \geq \max(S(A), S(B))$, in the separable case. So, in that sense, quantum separable correlations appear as classical. Based on the subadditivity, the non-negative quantity $I(A : B) = S(A) + S(B) - S(AB)$ accounts for the mutual information, or total correlations, between both parts.

For pure states, the entropy of the subsystems is *the* faithful measure of quantum correlations, under the name of "entropy of entanglement"; entanglement and discord coincide in that case. When dealing with mixed states, quantum correlations can also be measured by means of information-theoretic quantifiers but, as we have explained before, the notion lacks a univocal character. Entanglement measures are usually constructed by means of what is called "convex roof constructions of the pure states quantifiers," namely the entropy of entanglement, seeking for the optimal decomposition $\{p_i | \omega_i\rangle\}$ of the state ω into pure states $|\omega_i\rangle$. On the other hand, discord is computed as a difference between the mutual information of the state and the maximal mutual information accessible by measuring A and/or B independently.

Summary of Section 2

Many interesting results have been discussed in this section, many of them not usually made explicit in the standard presentation of quantum entanglement or other quantum correlations.

- The algebraic approach starts with an algebra of observables that are suitable to describe the system of interest. That *relevant algebra* characterizes the properties of our system, and the *state* of the system is represented by a functional that assigns a certain probability to each element of the algebra.
- There are states of maximal knowledge, the *pure* states, such that there is an observable that can be measured with certainty. Non-pure states are called "mixed" states.
- Two properties of a given system are *correlated* if the system is in a state such that the expectation of the product of observables is not the product of the marginal expectations.
- Given a pair of independent subsystems, we say that they are in a correlated state if there is at least one observable of each subsystem such that the pair is mutually correlated. Else, the system is in a *product* or uncorrelated state.
- *Entanglement* with respect to given subsystems implies the impossibility of expressing the state as a convex mixture of non-correlated states with respect to that subsystems.
- *Discord* with respect to given subsystems implies the impossibility of performing a local (with respect to that subsystems) measurement on the state without disturbing the joint state. Alternatively, discord implies the impossibility of writing the state as a convex mixture of local orthogonal projectors.
- Finally, both entanglement and discord are related to the non-commutative character of the quantum observables (i.e., to the non-abelian character of the corresponding algebras).

3 The Relative Nature of Quantum Correlations

In the previous section, we concluded that quantum correlations cannot be determined if there is no prior knowledge about the way in which our system is partitioned into subsystems. Such a link forces us to speak of a "relative" character of the correlations in the following sense. Suppose that we have two particles with spins one-half. The relevant algebras for each subsystem are the ones spanned by the local identities and the local spin operators, $\sigma_i \otimes \mathbb{I}$ and $\mathbb{I} \otimes \sigma_j$, where $\sigma_{i(j)}$, with $i,j = x, y, z$, are the Pauli matrices. Thus, the algebra for the composite system is spanned by the set $\mathcal{O} = \{\mathbb{I} \otimes \mathbb{I}, \sigma_i \otimes \mathbb{I}, \mathbb{I} \otimes \sigma_j \,|\, i, j = x, y, z\}$. This algebra induces the partition into mutually commutative subalgebras spanned by the sets of observables $\mathcal{A} = \{\mathbb{I} \otimes \mathbb{I}, \sigma_i \otimes \mathbb{I} \,|\, i = x, y, z\}$ and $\mathcal{B} = \{\mathbb{I} \otimes \mathbb{I}, \mathbb{I} \otimes \sigma_j \,|\, j = x, y, z\}$. These two sets represent, accordingly, two independent subsystems. But are they the only subsystems that one can identify for the system of two spins? The answer is an outright "no." As Zanardi and colleagues (2004) showed, the alternative set $\mathcal{O}' = \{\mathbb{I} \otimes \mathbb{I}, \mathbb{I} \otimes \mathbb{I}\sigma_x, \sigma_y \otimes \sigma_z, \sigma_z \otimes \sigma_z, \mathbb{I} \otimes \sigma_z, \sigma_x \otimes \sigma_y, \sigma_x \otimes \sigma_x\}$ induces a bipartition in terms of two subsystems that are different from the original ones, in the sense that they are not prescribed in terms of the local spin operators, namely $\mathcal{A}' = \{\mathbb{I} \otimes \mathbb{I}, \mathbb{I} \otimes \sigma_x, \sigma_y \otimes \sigma_z, \sigma_z \otimes \sigma_z\}$ and $\mathcal{B}' = \{\mathbb{I} \otimes \mathbb{I}, \mathbb{I} \otimes \sigma_z, \sigma_x \otimes \sigma_y, \sigma_x \otimes \sigma_x\}$. What is even more interesting, a state that is maximally entangled with respect to the first partition is non-correlated with respect to the second one.

Another typical example is the hydrogen atom, considered as a system of two particles, proton and electron. In order to find its ground state, it is convenient to define some virtual particles: the center of mass and the relative position particles. The ground state is (approximately) a non-correlated state of the virtual particles, while returning to the proton-electron description yields an entangled state (Tommasini et al. 1998).

One could ask whether one of those partitions, $\mathcal{O}$ or $\mathcal{O}'$, is more appropriate than the other, or has a more fundamental meaning. Zanardi and colleagues (2004) tackled the question in terms of the experimental accessible degrees of freedom, leading to the assertion that "quantum tensor product structures are observable-induced." They claim that, among all the possible decompositions of a subsystem into subsystems, the relevant ones are those consistent with the locally accessible degrees of freedom. However, there is much more to say about the many partitions of a system into subsystems and their impact on the way we determine the quantum correlations. For the moment, we are going to adopt a pluralist position, focusing on the variety of decompositions instead of a singular preferred one, postponing the debate about this possible ambiguity until Subsection 4.1.

In the following, we will show that the non-unique decomposition of a quantum system into subsystems is generic in the quantum composite systems' scenario. For that, we will consider an important result due to Harshman and Ranade (2011) about the entanglement of pure states. Moreover, we will discuss how to adopt a subsystem-independent view, first proposed by Barnum and colleagues (2003, 2004). We will rest our discussion on the enlightening analyses by Viola and Barnum (2010), Harshman (2012), and Earman (2014). Our hope is to motivate the idea that both the multiplicity of partitions into subsystems and the subsystem-independent perspective can be seen as inducing two stages in the analysis of the relative nature of quantum correlations. Before that, we need to understand the link between the usual notion of entanglement and the accessible knowledge in terms of the subsystems' observables.

3.1 Quantum Correlations and the Knowledge about Subsystems

As we said before, we can identify two ways in which the notion of correlation becomes relative in quantum mechanics. The first corresponds to a fact indicated previously: the decomposition of a system into subsystems can be effected in several different ways. Thus, the correlation is not a property of the state alone, but a property of the state-partition pair. For the example given earlier, suppose that the joint state *is a maximally entangled state, $|\psi\rangle$, with respect to $\Omega(\mathcal{O})$*. What does it imply to be maximally entangled with respect to that (relevant) algebra? As explained by Viola and Barnum (2010), from an operational point of view, the state $|\psi\rangle$ cannot be distinguished from a non-correlated state $|\psi^A\rangle \otimes |\psi^B\rangle$ *if we are restricted to measure local observables*. We are in the case of two spin one-half particles in terms of the algebra $\Omega(\mathcal{O})$ spanned by the set $\mathcal{O} = \{\mathbb{I} \otimes \mathbb{I}, \sigma_i \otimes \mathbb{I}, \mathbb{I} \otimes \sigma_j | i, j = x, y, z\}$. As stated before, this algebra induces the partition into mutually commutative subalgebras spanned by the sets of observables $\mathcal{A} = \{\mathbb{I} \otimes \mathbb{I}, \sigma_i \otimes \mathbb{I} | i = x, y, z\}$ and $\mathcal{B} = \{\mathbb{I} \otimes \mathbb{I}, \mathbb{I} \otimes \sigma_j | j = x, y, z\}$. Now, let us take $|\psi\rangle$ as a Bell-type state, $|\beta\rangle = (|\uparrow\rangle \otimes |\uparrow\rangle + |\downarrow\rangle \otimes |\downarrow\rangle)/\sqrt{2}$, that is maximally entangled. Computing only expectation values of $\mathcal{A}$ or $\mathcal{B}$ it is impossible to distinguish $|\beta\rangle$ from the non-entangled mixture $\rho_{\mathrm{mix}} = (|\uparrow\uparrow\rangle\langle\uparrow\uparrow| + |\downarrow\downarrow\rangle\langle\downarrow\downarrow|)/2$, where $|\uparrow\uparrow\rangle$ stands for $|\uparrow\rangle \otimes |\uparrow\rangle$. For example, the expectation value for $\mathbb{I} \otimes \sigma_z$ is Tr $(\rho_\beta \mathbb{I} \otimes \sigma_z) = \langle\beta|\mathbb{I} \otimes \sigma_z|\beta\rangle = 0 =$ Tr $(\rho_{\mathrm{mix}}\mathbb{I} \otimes \sigma_z)$ in both cases, and for $\mathbb{I} \otimes \sigma_x$ the result is Tr$(\rho_\beta \mathbb{I} \otimes \sigma_x) = 1/2 =$ Tr $(\rho_{\mathrm{mix}}\mathbb{I} \otimes \sigma_x)$. Thus, we can say that $|\beta\rangle$ and ρ_{mix} are $\Omega(\mathcal{O})$-indistinguishable. This effect is impossible if, instead of $|\beta\rangle$, we choose a non-entangled pure state as $|\psi\rangle = |\uparrow\rangle \otimes |\downarrow\rangle$. In the latter case, we can distinguish both states measuring, for example, $\mathbb{I} \otimes \sigma_z$, while it can be proved that there is no non-entangled mixture with respect to which $|\uparrow\rangle \otimes |\downarrow\rangle$ is $\Omega(\mathcal{O})$-is indistinguishable. Therefore, for pure states, $\Omega(\mathcal{O})$-distinguishability with respect to non-entangled mixtures seems an alternative way to characterize entanglement. Moreover, (non)entanglement for pure states can be characterized by noting that:

A pure state is non-entangled with respect to $\Omega(\mathcal{A}) \vee \Omega(\mathcal{B})$ if and only if it can be completely specified by the expectation values of the observables of the subsystems algebras $\Omega(\mathcal{A})$ and $\Omega(\mathcal{B})$.

This statement is straightforwardly proven when one writes the pure state in the Schmidt form: $|\psi\rangle = \sum_k \lambda_k |k^A\rangle \otimes |k^B\rangle$, with $\langle k^A|k'^A\rangle = \delta_{kk'} = \langle k^B|k'^B\rangle$ the Kronecker delta and where $\sum_k |\lambda_k|^2 = 1$. Remember that a pure state is non-entangled if and only if only one of the λ_k's is different from zero; take λ_0, for $k = 0$, as the corresponding non-zero value. Then, for any observable $O^A \otimes O^B$, the expectation value is written in the Schmidt basis as $\langle \psi | O^A \otimes O^B | \psi \rangle = \sum_{k,l} \lambda_k \lambda_l \langle k^A | O^A | l^A \rangle \langle k^B | O^B | l^B \rangle$. Therefore, a pure state s i non-entangled (with respect to $\Omega(\mathcal{A}) \vee \Omega(\mathcal{B})$) if and only if $\langle \psi | O^A \otimes O^B | \psi \rangle = \langle 0^A | O^A | 0^A \rangle \langle 0^B | O^B | 0^B \rangle$ for any observable $O^A \otimes O^B$. Taking the O^A's and O^B's to span a basis for the corresponding subsystems sets, the state can be completely specified relying on the expectations of observables of the subsystems.

Hence, there is a link between entanglement and incomplete knowledge in terms of the subsystems' observables: a non-entangled pure state can be completely specified by computing the expectation values of the subsystems' observables, while a pure entangled state needs the calculation of "correlated" expectation values. Intuitively, a state proportional to $|\uparrow\rangle \otimes |\uparrow\rangle + |\downarrow\rangle \otimes |\downarrow\rangle$ cannot be fully specified just by computing expectations of $\mathcal{A}$ and B independently, since the correlations between them must be considered. This observation indicates a relational character of quantum entanglement with respect to the subsystems and motivates the first stage in this process of generalization of the quantum correlations.

3.2 First Stage: Many Decompositions into Subsystems

The algebra $\Omega(\mathcal{O})$ considered so far for the two spins is not the only possible one. If, on the contrary, we consider the alternative algebra $\Omega(\mathcal{O}')$ as spanned by the set of observable operators $\mathcal{O}' = \{\mathbb{I} \otimes \mathbb{I}, \mathbb{I} \otimes \sigma_x, \sigma_y \otimes \sigma_z, \sigma_z \otimes \sigma_z, \mathbb{I} \otimes \sigma_z, \sigma_z \otimes \sigma_y, \sigma_x \otimes \sigma_x\}$, the induced subsystems are given by $\mathcal{A}' = \{\mathbb{I} \otimes \mathbb{I}, \mathbb{I} \otimes \sigma_x, \sigma_y \otimes \sigma_z, \sigma_z \otimes \sigma_z\}$ and $\mathcal{B}' = \{\mathbb{I} \otimes \mathbb{I}, \mathbb{I} \otimes \sigma_z, \sigma_x \otimes \sigma_y, \sigma_x \otimes \sigma_x\}$. Now, computing, for example, the expectation value of $\sigma_x \otimes \sigma_x$ yields the results $\mathrm{Tr}\left(\rho_\beta \sigma_x \otimes \sigma_x\right) = 1$ and $\mathrm{Tr}(\rho_{\mathrm{mix}} \sigma_x \otimes \sigma_x) = 0$, respectively, and we can conclude that $|\beta\rangle$ and ρ_{mix} are $\Omega(\mathcal{O}')$-*distinguishable* (indeed, perfectly distinguishable). The key step here is in the consideration of a different partition into subsystems: $\mathcal{A}|\mathcal{B} \leftrightarrow \mathcal{A}'|\mathcal{B}'$. Both partitions are legitimate, since the corresponding degrees of freedom for the first subsystem are associated to observables that commute with the ones of the second subsystem, in each case. Thus,

for any physical system there are, a priori, many (legitimate) partitions into subsystems, and correlations turn out to be relative to the chosen partition.

A subtle issue must be pointed out here: we are not tackling the issue of system/subsystems decompositions in a deep, philosophical manner, as we have not even discussed individuation in quantum theory (see, e.g., Hasse 2013 for a recent work on the individuation of physical systems in quantum theory). In this sense, our subjects are mostly physical: systems and subsystems, defined in terms of properties, measurements, and independence conditions, mathematically traduced into algebraic concepts.

This result is straightforwardly generalized to any pure state: for any pure state that is non-entangled with respect to some choice of subsystems, there exists another choice of subsystems such that the state is entangled with respect to them. Harshman and Ranade (2011) proved this for arbitrary unstructured Hilbert spaces (see the example that follows). Later, de la Torre and colleagues (2010) proved a very similar statement using a rather different approach.

Example: Unstructured Hilbert Spaces

The most radical example of the relative character of correlations arises when no restriction is imposed on the admissible observables, i.e., when one is allowed to surf the whole sea of possible observables. In that case, it is easy to see that notions like product (non-correlated) state, classically correlated state, or quantum-correlated state are meaningless without specification of the preferred setting.

A formal presentation of this problem, that sheds significant light, involves a generic pure state over an unstructured Hilbert space with non-prime dimension d (Harshman and Ranade 2011). It is possible to show that *for any pure state that is uncorrelated with respect to some bipartition, there is another partition with respect to which the state is maximally entangled*. Indeed, one is allowed to tailor the observables in order to fix a desired degree of entanglement for any pure state. The formal result is that "any state on a finite-dimensional Hilbert space can have arbitrary bipartite entanglement properties" (Harshman and Ranade 2011: 3). The key idea is that all Hilbert spaces with the same dimension are isomorphic. Therefore, there is an isometry that provides the mapping between a Hilbert space with some tensor product structure and any other Hilbert space with another tensor product structure. Equivalently, one can think that the Hilbert space remains untouched while applying a unitary transformation to its states. The transformed state reveals the entanglement of the original state with respect to the new tensor product structure. The only conditions imposed by Harshman and Ranade regards the non-prime dimension of the Hilbert space – to guarantee the possible bipartition – and a relation between the integer factors of this dimension and the Schmidt rank of the transformed state.

Summing up, if there is no restriction over the observable setting, then the relativeness is maximal, in the sense that every quantum state can be seen as uncorrelated or maximally correlated, depending on the preferred observables.

3.3 Second Stage: No Need for Subsystems

The second stage is probably not as natural as the first one, as it requires relaxing the usual subsystem-dependent perspective. Mathematically, subsystem-independent notions are needed for operations that are independent of the tensor product structures of the Hilbert spaces. The algebraic approach provides the right way to do this. The subsystem-independent formalism was proposed by Barnum and colleagues (2003). Balachandran and colleagues (2013a, 2013b) advanced the algebraic approach, replacing the usual partial trace operation with the idea of "state restriction to subalgebras." These concepts not only generalize the usual notion of entanglement, but also offer a consistent characterization of correlations in a wider range of models, including the identical particles-case, for which the notion of entanglement was always elusive (see Subsection 4.2).

After many years of debate, a clear and unique definition of quantum entanglement, covering also the identical particles setting, is still missing. Our previous discussion about the relative character of quantum correlations indicates that seeking an absolute definition is an obsolete question. Usually, entanglement is tackled from a state-based perspective, where the structure of the density operator seems to play the main role. On the other hand, from the relevant observables perspective, the relative aspect of quantum correlations immediately emerges together with a natural way to extend the traditional definitions. In this section, we discuss these issues and provide some examples that illustrate this generalized perspective.

To sketch the idea, let us return to the previous example for the two spins one-half. We discussed the algebra $\Omega(\mathcal{O})$, that induces the usual $\mathcal{A}|\mathcal{B}$ partition in terms of the Pauli spin operators. Given that partition, one could be interested in the easiest way to compute expectation values for the observables that involves properties of solely the $\mathcal{A}$ subsystem, as $\mathcal{O}^A \otimes \mathbb{I}^B$. In that case, knowledge of the joint state $|\psi\rangle$ is not required, as $\langle\psi|\mathcal{O}^A \otimes\mathbb{I}^B|\psi = \sum_{k,l}\lambda_k\lambda_l^*\langle k^A|\mathcal{O}^A|l^A\rangle\ \langle k^B|\mathbb{I}^B|l^B\rangle = \sum_{k,l}\lambda_k\lambda_l^*\langle k^A|\mathcal{O}^A|l^A\rangle\delta_{kl} = \sum_k|\lambda_k|^2\langle k^A|\mathcal{O}^A|k^A\rangle = \mathrm{Tr}_B(\mathcal{O}^A\rho^A)$, with $\rho^A := \sum_k|\lambda_k|^2|k^A\rangle\langle k^A| = \mathrm{Tr}_B|\psi\rangle\langle\psi|$, where " Tr_B" is the partial trace over the $\mathcal{B}$ degrees of freedom. Hence, instead of $|\psi\rangle$, knowledge of the state ρ^A, named "reduced state" for $\mathcal{A}$, is sufficient to compute expectation values for $\mathcal{O}^A$. But, as discussed in Section 3.1, only the non-correlated states can be completely specified by the expectation values of observables of the subsystems algebras $\Omega(\mathcal{A})$ and $\Omega(\mathcal{B})$. An immediate consequence of this is that *the reduced state is not pure unless the pure joint state is uncorrelated.* In other words, complete knowledge of the state of the system does not imply complete knowledge of the states of the subsystems, except in the case when the subsystems are in an uncorrelated state. This observation motivates another characterization of the pure entangled states (Balachandran 2013a, 2013b):

A pure state is non-entangled with respect to $\Omega(\mathcal{A}) \vee \Omega(\mathcal{B})$ if and only if the state restriction to the subalgebras is also pure.

The operation of *restriction to a subalgebra* is a natural algebraic generalization of the partial trace operation, based on the so-called GNS construction (see, e.g., Haag 2012 for details). The improvement with respect to the partial trace relies in that it is independent of any Hilbert space representation and, thus, it can be applied even when no tensor product structure of the Hilbert space is present. The novelty here is that an alternative characterization of entanglement is possible (at least, for pure states) that gets rid of the need to define a decomposition of the system into subsystems:

Definition 4 (Generalized entanglement). Let $\Omega(\mathcal{O})$ be the algebra of observables of a quantum system. A pure state ω over $\Omega(\mathcal{O})$ is non-entangled with respect to the subalgebra $\Omega(\mathcal{S}) \subseteq \Omega(\mathcal{O})$ if its restriction to the subalgebra, $\omega|_{\Omega(\mathcal{S})}$, is also pure. Else, it is entangled with respect to $\Omega(\mathcal{S})$.

This generalization of entanglement can be extended to mixed states using the idea of convex-roof construction, as explained by Barnum and colleagues (2004). When $\Omega(\mathcal{S})$ defines a subsystem algebra, one recover the usual setting and generalized entanglement becomes the usual entanglement. For pure states, generalized entanglement is clearly linked to the purity of the states, making the former a good candidate to quantify the generalized entanglement.

Example: Purity as a Measure of Generalized Entanglement

An interesting approach to generalize quantum entanglement is the one proposed by Barnum and colleagues (2003, 2005). Basically, their approach relies on the observation – pioneered by Schrödinger (1935) – that for a system in a quantum entangled state, the subsystems' properties are not well defined. This is what one usually discusses via the concept of *purity relative to a given subalgebra of relevant observables*. Maximal entanglement implies that expectations of subsystems' properties reveal total uncertainty: we cannot know the state of the whole by performing local measurements, and knowing the joint state says nothing about the state of the parts. We refer the reader to Zanardi and colleagues (2004) for a presentation of more detailed examples. As Girolami and colleagues (2013) explain, that intrinsic uncertainty is unavoidable, because a pure joint state is not an eigenstate of the subsystems' observables. Thus, any measurement corresponding to subsystems' observables disturbs the joint state. This observation can be rephrased in terms of a relative purity: a pure entangled state restricted to local observables exhibits mixedness (non-purity). The *relative purity under restricted capabilities* is an alternative characterization of quantum entanglement. Barnum and colleagues (2003, 2004, 2005) propose quantitative measures of generalized entanglement based on this fact.

A subsystems' state purity is then related to the usual notion of entanglement. Are both equivalent? The immediate answer is "no," because there are many situations in which the usual notion of entanglement (decomposability in terms of product states) is not well defined (such is the case for identical particles, which we are going to discuss later in Subsection 4.2). However, when the relevant observables that we select define a proper subsystem decomposition, generalized entanglement and (usual) entanglement are the same (Barnum et al. 2004; Viola et al. 2005; Viola and Barnum 2010).

3.4 Relative Character beyond Entanglement

So far, we have analyzed how the entanglement notion is relative to the non-unique substructure of the system and how a generalized notion of entanglement can be subsystem-independent. But this "relative-to-subsystems behavior" is also endemic for discord-like correlations. First, recall that non-discordant states are those for which there exists a subsystem's observable that can be measured without disturbing the joint state. As noted before, entanglement and discord coincide in the pure state scenario: the only states that are invariant under the non-selective measurement of a certain subsystem's observable are the pure uncorrelated ones. For mixed states, the coincidence breaks off. Instead, there emerges an inclusion relation: non-discordance (with respect to certain subalgebras) implies non-entanglement (with respect to the same subalgebras). Inversely, an entangled state is a discordant one. Let us discuss now the generalization of discord as we have earlier done with entanglement.

The first stage of generalization is analogous to that for entanglement. Considering different subsystems leads to different subalgebras. Non-disturbance with respect to one observable of the old subalgebras does not imply non-disturbance with respect to an observable of the new subalgebras. For example, for the two spins one-half system, the Bell state $|\beta\rangle$ is discordant with respect to the subalgebras $\Omega(\mathcal{A})$ and $\Omega(\mathcal{B})$ – since there is no observable on $\Omega(\mathcal{A})$ or $\Omega(\mathcal{B})$ that can be measured without disturbing the state – but non-discordant with respect to the subalgebras $\Omega(\mathcal{A}')$ and $\Omega(\mathcal{B}')$, since $|\beta\rangle$ is an eigenstate of $\sigma_z \otimes \sigma_z$. Hence, in the general case, *discord is relative to the partition of the system into subsystems.* In the pure state scenario, one could reproduce the radical relativeness results already proven for entanglement (see Subsection 3.2), since discord and entanglement are the same in those cases. For mixed states, the effect of considering different subsystems has not been explored in detail yet, so further analysis should reveal if the relative character is weakened when purity is not maximal.

The second stage of generalization depends on a subsystem-independent extension of the concept. *Our proposal* is that the notion of *relative coherence under restricted capabilities* underlies the concept of discord, as the notion of relative purity under restricted capabilities underlies the concept of entanglement. To understand our proposal, it is necessary to recognize the relation between measurement-disturbance and coherence. First, given a state ω with density operator ρ_ω over the Hilbert space $\mathcal{H}$, and for a fixed basis $K = \{|k\rangle\}$ of $\mathcal{H}$, *we say that ω is K-incoherent if $\rho_\omega = \sum_k p_k |k\rangle\langle k|$* (Baumgratz et al. 2014). As a consequence, the state is *K-incoherent* if it remains undisturbed under a non-selective measurement of any observable with spectrum K. Any quantum state is Λ-incoherent, where Λ is the spectrum of ρ_ω. However, things change when considering a restricted set of observables. For example, for a bipartite state that can be partitioned into subsystems $\mathcal{A}$ and $\mathcal{B}$, a state ρ_ω is K^A-incoherent if there is a basis $K^A = \{|k^A\rangle\}$ for the Hilbert space of $\mathcal{A}$ such that $\rho_\omega = \sum_{kl} p_{kl} |k^A\rangle\langle k^A| \otimes \rho_k^B$. Is it possible to give an informational measure of the coherence of a state with respect to a given observable? The question, tackled by several authors (see, e.g., Luo 2003; Girolami et al. 2013; Girolami 2014 and references therein), is answered affirmatively. A good option is to consider the Wigner-Yanase-Dyson skew information defined by Wigner and Yanase (1963): for a quantum state ρ_ω the K-coherence turns out to be (Girolami et al. 2013) $\mathcal{I}(\rho_\omega, K) = -\mathrm{Tr}([\sqrt{\rho_\omega}, K]^2)/2$. The quantity $\mathcal{I}(\rho_\omega, K)$ behaves as expected for a coherence monotone measure. Moreover, if one takes observables of one of the subsystems, as K^A from $\mathcal{A}$, Girolami showed that the corresponding minimal coherence, $\min_{K^A}\mathcal{I}(\rho_\omega, K^A)$, provides a bona fide discord-type measure of correlations. If one accepts that coherence with respect to subsystem observables is the property that underlies the quantum discord-like correlations, the natural subsystem-independent extension should involve the *coherence relative to a subalgebra*:

Definition 5 (Generalized discord). Let $\Omega(\mathcal{O})$ be the algebra of observables of a quantum system. A state ω over $\Omega(\mathcal{O})$ is (generalized) non-discordant with respect to the subalgebra $\Omega(\mathcal{S}) \subseteq \Omega(\mathcal{O})$ if there exists $S \in \Omega(\mathcal{S})$ with nondegenerate spectrum, such that ω is S-incoherent. Else, it is (generalized) discordant with respect to $\Omega(\mathcal{S})$.

This algebraic characterization of (non)discordant states is consistent with the usual measures of quantum discord, as all of them identify the same set of non-discordant states. The condition of nondegenerate spectrum is necessary, because we are interested in maximally informative observables; in other cases, the problem becomes trivial. Once more, the non-abelian character of the algebra of observables is essential, as it is evident from the measure $\mathcal{I}(\rho_\omega, K)$, which depends on the commutativity between the state, ρ_ω, and the observable K. Definition 5 offers a subsystem-independent extension of the notion of discord, in the same spirit as the generalization of entanglement proposed earlier, in Definition 4.

Further aspects of the relative character of entanglement and this algebraic approach have been studied by Thirring and colleagues (2011) and Derkacz and colleagues (2012), showing that the usual methodology based on the tensor product structure of the Hilbert spaces is unpleasantly limited.

So far, we have presented a discussion regarding the relational aspects between quantum correlations and the way in which a system is partitioned into subsystems or, even better, the observables that are taken into account for its description. What is still missing, and what we try to offer in Section 4, is a debate on the hierarchy (or not) of the multiplicity of possible decompositions into subsystems.

Summary of Section 3

We can summarize the conclusions of the present section as follows:

- In general, fixing the algebra of observables for one system does not prescribe a unique decomposition of the system into subsystems. There are multiple ways to decompose a quantum system into subsystems.
- The multiplicity of alternative decompositions into subsystems entails an intrinsic relative nature of quantum correlations, as the latter are determined with respect to the subsystems or, more generally, with respect to the subalgebras of relevant observables.
- In particular, entanglement can be generalized to a subsystem-independent algebraic definition: a pure non-entangled state with respect to a given subalgebra remains pure when restricted to that subalgebra. The idea is extended to mixed states via a convex roof construction.
- Discord has also a natural subsystem-independent generalization: a non-discordant state with respect to a given subalgebra shows zero quantum coherence with respect to some maximally informative observable of that subalgebra.

4 Toward a Unified View of Quantum Correlations

As presented in Section 3, purity and coherence under restricted observation – or, in other words, with respect to a given subalgebra of observables – are closely related to the notions of entanglement and discord, respectively. Also, for a given algebra of observables corresponding to one system, there are several possible compatible subalgebras. Hence, quantum correlations must refer to each subalgebra. Is there any way to privilege one setting among all the others? If the answer is affirmative, then we have a criterion to determine a unique value of entanglement/discord for a certain state. Else, we must accept the diversity of settings and verify that there is no ambiguity at all.

4.1 Relevant Observables: In the Quest for a Preferred Description?

When looking for a privileged description – a preferred set of relevant observables with respect to which quantum correlations could be accounted for – it is useful to distinguish some kind of epistemic view from a more ontological one, or, as Earman (2014) puts it, to frame the issue in some sort of pragmatist versus realist debate.

Let us start by recalling Zanardi's position (2001). The fact of considering all the possible partitions on the same footing (that amounts to establishing a *democracy* between different tensor product structures) provides a relativization of the notion of entanglement. Without further physical assumption, no partition has an ontologically superior status with respect to any other. Considering a given partition as the privileged one has a strong operational meaning, in that the partition depends on the set of resources effectively available to access and to control the degrees of freedom of the system. Summarizing, Zanardi holds that experimental accessibility dictates the way in which a privileged partition must be selected. We find two objections to this view. First, experimental accessibility is not enough to select *a unique* partition. Second, *the set of resources effectively available* to access and to control the system cannot provide a universal principle, as they are strongly dependent on the current technics and technologies. What is more, the latter is an objection that applies to any pragmatist-like viewpoint. If one is decided to find a fundamental criterion to establish a hierarchy, we should look at another place.

Harshman and Wickramasekara's (2007a, 2007b) proposal, based on the symmetries and dynamical invariances of the system, constitutes a very different alterative. Considering symmetry-invariant tensor product structures gives place to measures of quantum correlations that do not depend on the frame of reference. In the same vein, but with a rather different idea, Harshman holds that entanglement comes before, in the sense that entanglement dictates how to decompose our system (2012: 386): "the physically-meaningful observable subalgebras are the ones that minimize entanglement in typical states." We think that, although not conclusive, these notions deserve much attention, as they offer physical motivated criteria to distinguish the variety of decompositions or, paraphrasing Zanardi, turning the partitions' democracy into an aristocracy where symmetry principles govern.

An important question refers to the *relevant observables* versus *locality* dichotomy. As the relevant observables are not generally conditioned to represent local accessible degrees of freedom, the distinction between relevant and local must be emphasized. The notion of entanglement is closely linked to the idea of local creation of correlations: under the Local Operations and Classical Communication (LOCC) paradigm (see Bennett et al. 1996; Bruß 1999), entanglement is always non-increasing after the performance of that kind of operations. Splitting entanglement from locality seems a bad idea if one wants to preserve that paradigm alive, and one may wonder if a generalization of entanglement has any sense at all if the prescription of locality is diluted. A theory is supposed to admit a local model if, among other conditions, results obtained at one location are independent of any actions performed at space-like separations (Bell 1964). Although locality conditions are not usually made explicit in the quantum information literature, one could assume that any multipartite setting is traditionally determined – in accordance to the pragmatic issues discussed earlier – in terms of locally accessible degrees of freedom. In the case that each part is space-like separated from the others, "mixing" their observables dilutes any idea of locality. A possible solution to this physical puzzle is, again, to consider not all the possible decompositions of our system on an equal footing, but only the ones *compatible* with the locality conditions. Thus, one could divide the variety of decompositions into two groups: those compatible with certain locality prescriptions (i.e., decomposition corresponding to subalgebras of time-like separated observables), and those *incompatible* with those prescriptions. However, there are non-local Hamiltonians and, therefore, non-local quantum computational gates (see, e.g., Vidal et al. 2002), and one should not discard the incompatible decompositions at all from a fundamental point of view. In other words, accepting an ultimate non-locality of quantum theory implies the consideration of non-local observables as legitimate as local ones (although not necessarily on an equal footing). What one should always keep in mind is that entanglement and non-locality are two different features of quantum systems (refer, e.g., to the typical works by Popescu 1994 and Bennett et al. 1999, or the more recent discussion by Brunner et al. 2005).

Summing up, although alternative descriptions in terms of different relevant observables are equivalent (in the sense that all of them, by construction, prescribe the same probabilities for any observable that one can measure), there may be a privileged description that is preferable due to some physically grounded argument (e.g., fundamental symmetries of the theory and/or model) or because of its computational simplicity. Anyway, consideration of relevant observables is a powerful approach: it is very general and fits nicely with the idea of correlations being relational concepts. Also, it is an objective view since it does not appeal to any kind of preferred observer involving its subjective perception. On the contrary, the relevant observables framework accepts that entanglement (and other correlations) are relative to the properties and cannot be independently ascribed to the quantum states.

4.2 *Identical Particles and Quantum Correlations*

The study of quantum correlations in identical particles is, for our present purposes, paradigmatic. Thus, although not delving into in the most technical details, we should mention some conceptual issues.

What is essential in this case is the situation where two or more particles are quantum mechanically indistinguishable – which is not precisely the same as being identical – meaning that there is permutation invariance. From the algebraic viewpoint, invoking this permutation invariance implies a super-selection rule, splitting the total algebra (of the system of indistinguishable particles) into two sectors, namely that spanned by symmetric (bosonic) and that spanned by antisymmetric (fermionic) vectors (see, e.g., Messiah and Greenberg 1964). If, for a system of N fermions, one begins from the single-particle Hilbert spaces, $\mathcal{H}_1$, the Hilbert space for the system is then the totally anti-symmetric space $\wedge \mathcal{H}_1^{\otimes N} \subset \mathcal{H}_1^{\otimes N}$, where $\wedge$ denotes the anti-symmetrization. This Hilbert space does not admit a tensor product structure in terms of the single-particle spaces, just as expected from the indistinguishability condition. Hence, one cannot identify proper subsystems in terms of which entanglement (or any other kind of correlation) could be formulated. Moreover, the partial trace operation is ill-defined in this case.

In order to solve this problem, different proposals arise. On one hand, one could ignore this algebraic matter at all, and make use of the usual measures of quantum entanglement of distinguishable systems. On the other hand, one could reformulate the problem in terms of a Hilbert space that admits a tensor product structure: the associated Fock space (Fock 1932). The Fock space is the configuration in terms of the number of fermions – zero or one – in each possible single-particle state. As Earman (2014) explains, getting married with this position implies some kind of pragmatic selection of preferred observables.

Again, the subsystem-independent generalizations of entanglement proposed by Barnum and colleagues (2003, 2004, 2005) can be suitably applied to indistinguishable particles, whenever a subspace of preferred observables can be determined. Balachandran and colleagues (2013a, 2013b) explore further this approach, explaining how the usual notion of partial trace is a particular case of *restriction of a state to a subalgebra*, where the latter is universally valid and does not require identification of subsystems. Finally, it is interesting to recall the works by Benatti and colleagues (2010, 2014a, 2014b), where the authors study the entanglement of identical particles in this spirit of a preferred observables setting, with special focus on the idea of algebraic independence between different sets of observables.

Summary of Section 4

- The multiplicity of subsystems decompositions – or subspaces of preferred relevant observables – points toward two mutually exclusive scenarios: either (a) one defines a criterion for selecting a uniquely privileged setting, or (b) one accepts the plurality of configurations and shows that this does not imply ambiguity for the notion of correlations.
- With regards to the viewpoint (a), the criteria found in the literature are linked to experimental accessibility, locality, or dynamical invariance of entanglement under symmetry transformations.
- Nonetheless, giving the implausibility that one criterion could select a uniquely preferred configuration, choosing the viewpoint (b) seems to be the right procedure. Ambiguity is avoided if correlations are understood as relational between the relevant observables and the quantum state.
- Criteria like those used in viewpoint (a) can be used to define some kind of physically based hierarchy in the catalog of possible decompositions, discerning between subalgebras of locally compatible observables, symmetry-invariant ones, etc.
- Indistinguishability provides an interesting testbed for the generalized (subsystem-independent) notion of quantum correlations.

5 Classical Aspects of Quantum Mechanics

In Subsection 3.4, we introduced the notion of general quantum correlations that go beyond entanglement, as discord-like measures. There is a remarkable difference between entanglement and general quantum correlations – a difference that is very important for our discussion – a non-entangled state with respect to certain bipartition remains non-entangled with respect to any other bipartition *locally compatible* with the original one. That is not true for general quantum correlations. As a consequence, we can show how the consideration of different relevant sets of observables change the correlation-perspective of a given state.

5.1 Classical Features from Quantum-Correlated States

When prescribing a partition of a system into subsystems, one is determining a preferred set of observables to describe the system. No matter what the criterion to define the preference is, in general there are many other options that are also compatible but lead to different results concerning the correlations scheme. We argue that, as correlations are relational concepts between the state and the observables that define the system, the quantumness (or classicality) of the correlations exhibits also a relational status. That is, a quantum system exhibits quantum correlations with respect to almost every pair of observables but, in some cases, the correlations display some classical properties. However, for classically correlated states, there exist other observables for which quantum features are unavoidable.

There is a "hidden quantumness" for classically correlated states that plays a key role in the quantum information scenario. Let us consider the following scenario: a classically correlated state ρ_{class}^{AB} in $\mathcal{H}^A \otimes \mathcal{H}^B$, such that the dimension of A is a non-prime integer. In this case, *A can be* considered as a bipartite system, $A \cong A_1 | A_2$, and it is valid to ask about the correlations between A_1 (or A_2) and B. In other words, we are concerned with the reduced state $\rho^{A_1 B} = \mathrm{Tr}_{A_2} \rho_{class}^{AB}$. It turns out that, in general, $\rho^{A_1 B}$ is not a classically correlated state: A *classically correlated state with respect to $A|B$ is generally quantum-correlated with respect to $A_1|B$*. Li and Luo (2008) first emphasized this point, showing that any separable state is (formally) a reduction of a classically correlated state in higher dimensions, preserving the natural bipartition. They argue that separable states can be regarded as "natural shadows" of classical states in larger systems. Many interesting results regarding quantum information processing follow from such observation due to Li and Luo. First, it is straightforward to prove that any quantum task that does not need entanglement can be carried out starting with classically correlated states as the unique resource (Bellomo et al. 2014, 2015, 2016). In turn, classically correlated (quantum) states reveal themselves as a legitimate quantum resource (Plastino et al. 2015).

Although this "quantum-out-from-classical" behavior seems to contradict the title of the current subsection – namely, "Classical features from quantum-correlated states" – this is not the case. What we show here is that *some types of quantum correlations can appear as classical correlations when certain relevant observables in a larger Hilbert space are measured*. Given the non-abelian character of the algebras of quantum observables, it is always possible to turn to an alternative description in terms of different observables where the quantum nature of the correlations is unhidden.

Summing up, although the information-theoretic view of quantum theory brings a novel perspective and useful tools for defining quantifiers for peculiar quantum correlations, it is hard to believe that in this way one could understand the transition between a quantum reality and a classical one (see Subsection 5.3). As Devi and Rajagopal (2008: 1) contend, "the notion of correlation *per se* does not set a borderline between classical and quantum descriptions."

5.2 Classical Features of Quantumness

Let us suppose that we find a relevant set of observables such that, in their terms, a certain quantum system is classically correlated. As previously stated, this classical character is never absolute because it depends on the preferred observables that one selects. Nonetheless, we can state that such a system *appears* to be a classical system *with respect to* a relevant set of observables. Different measurements that are not compatible with this relevant set would reveal quantum features, but if one remains within the limits of those preferred measurements – that is, when computing expectations for the preferred observables – the quantum nature will remain hidden from observation.

A remarkable consequence stems in the understanding of the decoherence process. Decoherence has been the mainstream explanation of the transition between the quantum and the classical reality. Introduced by Zeh in the early 1970s (e.g., 1970), and with essential contributions of Zurek (1981, 1982), decoherence basically attempts to provide a quantum mechanical process to substantiate the unobservable quantum nature of the macroscopic reality and, as it is closely related with the former, to resolve the measurement problem. As Schlosshauer (2007) prescribes, the measurement problem involves three questions: (a) the problem of the preferred basis, namely, the way in which preferred physical quantities appear in nature (e.g., why do we observe, in many cases, definite positions rather than superpositions of positions?); (b) the non-observability of interference, mainly in the macroscopic scenario; (c) the fact that, besides the probabilistic character of the quantum formalism, each measurement has a definite outcome (what selects the corresponding outcomes?). It is commonly accepted that decoherence brings a possible solution for the first two problems, while question (c) remains unanswered (see, e.g., Wallace 2012). Regarding (a)–(b) and its relation to our present discussion, it must be stressed that decoherence attempts to solve (a) by selecting as preferred the basis associated to the dynamical interaction between the system and its environment, and attempts to solve (b) by showing that a high-dimensional environment leads to a virtually unobservable interference in the mentioned basis. The already decohered state of the system exhibits zero discord (see Definition 3), since it is diagonal in the preferred basis. It is straightforward to see, now, that the relative character of quantum discord can be linked to a relative occurrence of the decoherence process (Castagnino et al. 2008, 2009, 2010a, 2010b; Lombardi et al. 2011, 2012). A quantitative model displaying the relation between decoherence and the way a system is partitioned has been presented by Lychkovskiy (2013). To be fair, we should mention that Zurek himself had pointed out this difficulty (1998: 1820): "one issue which has been often taken for granted is looming big, as a foundation of the whole decoherence program. It is the question of what are the 'systems' which play such a crucial role in all the discussions of the emergent classicality."

As stated before, regarding the third question – item (c) – the possible role of the decoherence mechanism seems unsubstantiated. As Schlosshauer says (2007: 50): "the success of decoherence in tackling the third issue – the problem of outcomes – remains a matter of debate, in particular, because this issue is almost inextricably linked to the choice of a specific interpretation of quantum mechanics. ... In fact, most of the overly optimistic or pessimistic statements about the ability of decoherence to solve 'the' measurement problem can be traced back to a misunderstanding of the ability of a standard quantum effect such as decoherence may display in solving the more interpretive problem of outcomes." Different approaches to tackle this problem have been made and discussed from the viewpoint of Everett's and Modal interpretations (see, e.g., Wallace 2003; Lombardi 2011 and references therein; Bacciagaluppi 2012). Finally, it is noteworthy that Jeknić-Dugić and colleagues (2013) have posed an interesting problem concerning the compatibility of decoherence with many-worlds interpretations, and the relative character of quantum correlations.

Summary of Section 5

Regarding the troubling relation between classical and quantum we can conclude that:

- Classically correlated states are still quantum states: they are classically correlated with respect to given preferred subalgebras of observables, but turning to another preferred description (and due to the non-abelian character of the algebras of quantum observables) generally reveals the quantum nature of the states.
- As a consequence, classically correlated states are compatible with quantum-correlated reductions.
- The relative character of quantum correlations imposes restrictions on the classical appearance of the quantum: the classical appearance is also relative to a given preferred description, underlying a quantum character with respect to most of the other possible descriptions.

6 Concluding Remarks

Throughout this chapter, we have explored some aspects of quantum correlations from an algebraic and information-theoretic point of view, upholding a relative-to-the-relevant-observables perspective. This perspective has induced generalized definitions for the usual notions of quantum correlations. In turn, its subsystem-independent character has allowed for a consistent application of the generalized notions to every quantum mechanical scenario. Finally, we have discussed how the quantum/classical border is interpreted from this viewpoint. In summary, this relative characterization of quantum correlations should be taken seriously and studied in detail, as they could bring better understanding of some conceptual and pragmatic issues.

Acknowledgments

We are grateful to the participants of the International Workshop "What Is Quantum Information?", held in Buenos Aires on May 2015, with whom we could share and exchange many ideas and stimulating conversations.

References

Bacciagaluppi, G. (2012). "The Role of Decoherence in Quantum Mechanics." In E. N. Zalta (ed.), *The Stanford Encyclopedia of Philosophy*, http://plato.stanford .edu/archives/win2012/entries/qm-decoherence/.

Balachandran, A. P., Govindarajan, T. R., de Queiroz, A. R., and Reyes-Lega, A. F. (2013a). "Algebraic Approach to Entanglement and Entropy." *Physical Review A*, **88**: 022301.

Balachandran, A. P., Govindarajan, T. R., de Queiroz, A. R., and Reyes-Lega, A. F. (2013b). "Entanglement and Particle Identity: A Unifying Approach." *Physical Review Letters*, **110**: 080503.

Barnum, H., Knill, E., Ortiz, G., Somma, R., and Viola, L. (2003). "Generalizations of Entanglement Based on Coherent States and Convex Sets." *Physical Review A*, **68**: 032308.

Barnum, H., Knill, E., Ortiz, G., Somma, R., and Viola, L. (2004). "A Subsystem-Independent Generalization of Entanglement." *Physical Review Letters*, **92**: 107902.

Barnum, H., Ortiz, G., Somma, R., and Viola, L. (2005). "A Generalization of Entanglement to Convex Operational Theories: Entanglement Relative to a Subspace of Observables." *International Journal of Theoretical Physics*, **44**: 2127.

Baumgratz, T., Cramer, M., and Plenio, M. B. (2014). "Quantifying Coherence." *Physical Review Letters*, **113**: 140401.

Bell, J. S. (1964). "On the Einstein-Podolsky-Rosen Paradox." *Physics*, **1**: 195–200.

Bellomo, G., Majtey, A. P., Plastino, A. R., and Plastino, A. (2014). "Quantum Correlations from Classically-Correlated States." *Physica A: Statistical Mechanics and Its Applications*, **405**: 260–266.

Bellomo, G., Plastino, A., and Plastino, A. R. (2015). "Classical Extension of Quantum-Correlated Separable States." *International Journal of Quantum Information*, **13**: 1550015.

Bellomo, G., Plastino, A., and Plastino, A. R. (2016). "Quantumness and the role of locality on quantum correlations." *Physical Review A*, **93**: 062322.

Benatti, F., Floreanini, R., and Marzolino, U. (2010). "Sub-Shot-Noise Quantum Metrology with Entangled Identical Particles." *Annals of Physics*, **325**: 924–935.

Benatti, F., Floreanini, R., and Marzolino, U. (2014a). "Entanglement in Fermion Systems and Quantum Metrology." *Physical Review A*, **89**: 032326.

Benatti, F., Floreanini, R., and Titimbo, K. (2014b). "Entanglement of Identical Particles." *Open Systems & Information Dynamics*, **21**: 1440003.

Bengtsson, I. and Zyczkowski, K. (2006). *Geometry of Quantum States: An Introduction to Quantum Entanglement*. Cambridge: Cambridge University Press.

Bennett, C. H., Bernstein, H. J., Popescu, S., and Schumacher, B. (1996). "Concentrating Partial Entanglement by Local Operations." *Physical Review A*, **53**: 2046.

Bennett, C. H., Di Vincenzo, D. P., Fuchs, C. A., Mor, T., Rains, E., Shor, P. W., and Wootters, W. K. (1999). "Quantum Nonlocality without Entanglement." *Physical Review A*, **59**: 1070.

Born, M. (1926). "Quantenmechanik der Stoßvorgänge." *Zeitschrift für Physik*, **38**: 803–827.

Bratteli, O. and Robinson, D. W. (2012). *Operator Algebras and Quantum Statistical Mechanics: Volume 1: C*-and W*-Algebras. Symmetry Groups. Decomposition of States*. Berlin: Springer-Verlag.

Brunner, N., Gisin, N., and Scarani, V. (2005). "Entanglement and Non-locality Are Different Resources." *New Journal of Physics*, **7**: 88.

Bruß, D. (1999). "Entanglement Splitting of Pure Bipartite Quantum States." *Physical Review A*, **60**: 4344.

Castagnino, M., Fortin, S., Laura, R., and Lombardi, O. (2008). "A General Theoretical Framework for Decoherence in Open and Closed Systems." *Classical and Quantum Gravity*, **25**: 154002.

Castagnino, M., Fortin, S., and Lombardi, O. (2009). "Decoherence as a Relative Phenomenon: A Generalization of the Spin-Bath Model." *arXiv preprint*, arXiv:0907.1933.

Castagnino, M., Fortin, S., and Lombardi, O. (2010a). "Suppression of Decoherence in a Generalization of the Spin-Bath Model." **43**: 065304.

Castagnino, M., Fortin, S., and Lombardi, O. (2010b). "Is the Decoherence of a System the Result of Its Interaction with the Environment?" *Modern Physics Letters A*, **25**: 1431–1439.

De la Torre, A. C., Goyeneche, D., and Leitao, L. (2010). "Entanglement for All Quantum States." *European Journal of Physics*, **31**: 325–332.

Derkacz, Ł., Gwóźdź, M., and Jakóbczyk, L. (2012). "Entanglement beyond Tensor Product Structure: Algebraic Aspects of Quantum Non-separability." *Journal of Physics A: Mathematical and Theoretical*, **45**: 025302.

Devi, A. U. and Rajagopal, A. K. (2008). "Generalized Information Theoretic Measure to Discern the Quantumness of Correlations." *Physical Review Letters*, **100**: 140502.

Earman, J. (2014). "Some Puzzles and Unresolved Issues about Quantum Entanglement." *Erkenntnis*, **80**: 303–337.

Fock, V. (1932). "Konfigurationsraum und Zweite Quantelung." *Zeitschrift für Physik*, **75**: 622–647.

Girolami, D. (2014). "Observable Measure of Quantum Coherence in Finite Dimensional Systems." *Physical Review Letters*, **113**: 170401.

Girolami, D., Tufarelli, T., and Adesso, G. (2013). "Characterizing Nonclassical Correlations via Local Quantum Uncertainty." *Physical Review Letters*, **110**: 240402.

Haag, R. (2012). *Local Quantum Physics: Fields, Particles, Algebras*. Berlin: Springer.

Hamhalter, J. (2003). "Generalized Gleason Theorem." Pp. 121–175 in *Quantum Measure Theory*. Dordrecht: Springer.

Harshman, N. L. (2012). "Observables and Entanglement in the Two-Body System." Pp. 386–390 in *AIP Conference Proceedings, Quantum Theory: Reconsideration of Foundations 6*, Linnaeus University, June 11–14.

Harshman, N. L. and Ranade, K. S. (2011). "Observables Can Be Tailored to Change the Entanglement of Any Pure State." *Physical Review A*, **84**: 012303.

Harshman, N. L. and Wickramasekara, S. (2007a). "Galilean and Dynamical Invariance of Entanglement in Particle Scattering." *Physical Review Letters*, **98**: 080406.

Harshman, N. L. and Wickramasekara, S. (2007b). "Tensor Product Structures, Entanglement, and Particle Scattering." *Open Systems & Information Dynamics*, **14**: 341–351.

Hasse, C. L. (2013). "On the Individuation of Physical Systems in Quantum Theory." PhD Thesis, University of Adelaide.

Henderson, L. and Vedral, V. (2001). "Classical, Quantum and Total Correlations." *Journal of Physics A: Mathematical and General*, **34**: 6899–6905.

Horodecki, R., Horodecki, P., Horodecki, M., and Horodecki, K. (2009). "Quantum Entanglement." *Reviews of Modern Physics*, **81**: 865–942.

Jeknić-Dugić, J., Arsenijević, M., and Dugić, M. (2013). *Quantum Structures: A View of the Quantum World*. Saarbrücken: Lambert Academic Publishing.

Li, N. and Luo, S. (2008). "Classical States versus Separable States." *Physical Review A*, **78**: 024303.

Lombardi, O., Ardenghi, J. S., Fortin, S., and Castagnino, M. (2011). "Compatibility between Environment-Induced Decoherence and the Modal Hamiltonian Interpretation of Quantum Mechanics." *Philosophy of Science*, **78**: 1024–1036.

Lombardi, O., Fortin, S., and Castagnino, M. (2012). "The Problem of Identifying the System and the Environment in the Phenomenon of Decoherence." In H. W. de Regt, S. Hartmann, and S. Okasha (eds.), *EPSA Philosophy of Science: Amsterdam 2009*. Berlin: Springer.

Luo, S. (2003). "Wigner-Yanase Skew Information and Uncertainty Relations." *Physical Review Letters*, **91**: 180403.

Lychkovskiy, O. (2013). "Dependence of Decoherence-Assisted Classicality on the Way a System Is Partitioned into Subsystems." *Physical Review A*, **87**: 022112.

Messiah, A. M. L. and Greenberg, O. W. (1964). "Symmetrization Postulate and Its Experimental Foundation." *Physical Review*, **136**: B248–B267.

Modi, K., Brodutch, A., Cable, H., Paterek, T., and Vedral, V. (2012). "The Classical-Quantum Boundary for Correlations: Discord and Related Measures." *Reviews of Modern Physics*, **84**: 1655–1707.

Nielsen, M. A. and Chuang I. L. (2010). *Quantum Computation and Quantum Information*. Cambridge: Cambridge University Press.

Ollivier, H. and Zurek, W. H. (2001). "Quantum Discord: A Measure of the Quantumness of Correlations." *Physical Review Letters*, **88**: 017901.

Plastino, A., Bellomo, G., and Plastino, A. R. (2015). "Quantum State Space-Dimension as a Quantum Resource." *International Journal of Quantum Information*, **13**: 1550039.

Popescu, S. (1994). "Bell's Inequalities versus Teleportation: What Is Nonlocality?" *Physical Review Letters*, **72**: 797–799.

Raggio, G. A. (1988). "A Remark on Bell's Inequality and Decomposable Normal States." *Letters in Mathematical Physics*, **15**: 27–29.

Rédei, M. (2013). *Quantum Logic in Algebraic Approach*. Dordrecht: Springer.

Schlosshauer, M. A. (2007). *Decoherence and the Quantum-to-Classical Transition*. Dordrecht: Springer.

Schrödinger, E. (1935). "Discussion of Probability Relations between Separated Systems." *Mathematical Proceedings of the Cambridge Philosophical Society*, **31**: 555–563.

Thirring, W., Bertlmann, R. A., Köhler, P., and Narnhofer, H. (2011). "Entanglement or Separability: The Choice of How to Factorize the Algebra of a Density Matrix." *The European Physical Journal D*, **64**: 181–196.

Tommasini, P., Timmermans, E., and de Toledo Piza, A. (1998). "The Hydrogen Atom as an Entangled Electron-Proton System." *American Journal of Physics*, **66**: 881–886.

Vidal, G., Hammerer, K., and Cirac, J. I. (2002). "Interaction Cost of Nonlocal Gates." *Physical Review Letters*, **88**: 237902.

Viola, L. and Barnum, H. (2010). "Entanglement and Subsystems, Entanglement beyond Subsystems and All That." Pp. 16–43 in A. Bokulich and G. Jaeger (eds.), *Philosophy of Quantum Information and Entanglement*. Cambridge: Cambridge University Press.

Viola, L., Barnum, H., Knill, E., Ortiz, G., and Somma, R. (2005). "Entanglement beyond Subsystems." *Contemporary Mathematics*, **381**: 117.

Wallace, D. (2003). "Everett and Structure." *Studies in History and Philosophy of Modern Physics*, **34**: 87–105.

Wallace, D. (2012). "Decoherence and Its Role in the Modern Measurement Problem." *Philosophical Transactions of the Royal Society A*, **370**: 4576–4593.

Werner, W. F. (1989). "Quantum States with Einstein-Podolsky-Rosen Correlations Admitting a Hidden Variable Model." *Physical Review A*, **40**: 4277–4281.

Wigner, E. P. and Yanase, M. M. (1963). "Information Contents of Distributions." *Proceedings of the National Academy of Sciences of the United States of America*, **49**: 910–918.

Zanardi, P. (2001). "Virtual Quantum Subsystems." *Physical Review Letters*, **87**: 077901.

Zanardi, P., Lidar, D. A., and Lloyd, S. (2004). "Quantum Tensor Product Structures Are Observable Induced." *Physical Review Letters*, **92**: 060402.

Zeh, D. H. (1970). "On the Interpretation of Measurement in Quantum Theory." *Foundations of Physics*, **1**: 69–76.

Zurek W. H. (1981). "Pointer Basis of Quantum Apparatus: Into what Mixture Does the Wave Packet Collapse?" *Physical Review D*, **24**: 1516–1525.

Zurek W. H. (1982). "Environment-Induced Superselection Rules." *Physical Review D*, **26**: 1862–1880.

Zurek W. H. (1998). "Decoherence, Einselection and the Existential Interpretation (The Rough Guide)." *Philosophical Transactions of the Royal Society of London, Series A: Mathematical Physical and Engineering Sciences*, **356**: 1793–1821.

Index